FORESTRY IN NEW ZEALAND

FORESTRY IN NEW ZEALAND

The Shaping of Policy

A. L. POOLE
B.(For.)Sc., M.Sc., F.R.S.N.Z.
Director-General, N.Z. Forest Service

HODDER & STOUGHTON
AUCKLAND
in association with
THE ENGLISH UNIVERSITIES PRESS LTD.
LONDON

© A. L. POOLE 1969
First printed 1969
SBN 340 09902 X

Printed and bound in New Zealand for
Hodder & Stoughton Ltd. (Incorporated in England)
52 Cook Street, Auckland
by Wright & Carman Ltd., Wellington

CONTENTS

			Page
Chapter	1.	What is Forestry?	1
„	2.	A Forest Policy Evolves	3
„	3.	Native Forests and the Tradition of Wood Use	18
„	4.	Forests and Land Use	29
„	5.	The Development of Forest Management	49
„	6.	The Forest Products Industries	75
„	7.	Looking Ahead	86
„	8.	Training and Research	99
Appendix	1.	The principal New Zealand Acts affecting forestry	110
„	2.	Timber trees named in the text	111

LIST OF ILLUSTRATIONS

following page

1. A Maori *pa* in old Taranaki — 6
2. Early Nelson settlement — 6
3. Pitsawing kauri logs — 22
4. Government Buildings, Wellington — 22
5. Waipa State Mill — 30
6. Protection forest in Westland — 30
7. A rata-tawa forest undamaged by animals — 30
8. Erosion on the Craigieburn Range — 30
9. Part of the Naseby State Forest, Central Otago — 38
10. Golden Downs State Forest, Nelson — 38
11. Extensive erosion in the Poverty Bay-East Coast region — 46
12. Advanced slumping at Waipaoa, Poverty Bay — 46
13. Second growth at Te Wera, Taranaki — 46
14. Sand dune reclamation with radiata pine on the Wellington west coast — 46
15. Kahikatea forest in South Westland — 54
16. Rimu terrace forest in Westland — 54
17. Regenerating rimu forest on the West Coast — 54
18. Kauri in Russell State Forest — 54
19. Mature kauri in Waipoua Forest Sanctuary — 62
20. Slow-growing New Zealand cedar — 62
21. Conical Hill, Southland — 62
22. An untended 35-year-old stand of radiata pine in Kaingaroa Forest — 62

LIST OF ILLUSTRATIONS

following page

23. High quality radiata pine on a Wairarapa farm — 70
24. Natural regeneration of radiata pine after fire — 70
25. Natural regeneration thinned and low pruned — 70
26. Douglas fir at Mount Peel, Canterbury — 78
27. Corsican pine in Kaingaroa Forest — 78
28. 40-year-old lodgepole pine in Hanmer Forest — 78
29. Lodgepole pine regeneration on Mount Ruapehu — 78
30. An early sawmill near Wellington — 86
31. Kauri logs floating to the mill — 86
32. A sawmill cutting native timber — 102
33. Japanese ships loading pine logs at Mount Maunganui — 102

ACKNOWLEDGEMENTS

Alexander Turnbull Library 1, 2, 3, 4, 30.

Dominion Museum 7, 8.

New Zealand Forest Service, J. H. Johns 6, 9, 10, 11, 12, 13, 14, 15, 17, 23, 25, 26, 27, 28, 29, 32.

New Zealand Forest Service 5, 21, 22, 33.

New Zealand Forest Service, W. J. Wilson 18, 19, 20.

Weekly News 31.

D.S.I.R. 16.

AUTHOR'S NOTE

The introduction to this book emphasises the many-sided nature of forest management and forest policy. Therefore, in writing about New Zealand developments over a period of more than 100 years, many sources of information have been consulted.

The most important have been parliamentary and official papers, documents and files, because it is Government that is primarily responsible for moulding forest policy. Governments are vitally concerned with forests and land use and with forests and timber use. A few of the references used are given throughout the text but no comprehensive list has been compiled. To the reader wanting a more complete background to statements made, the following references are some of the more important additions to those given:

- N.Z. Department of Lands and Survey. Annual Reports, 1877/78–1917/18.
- N.Z. Forest Service (1918/19 Department of Forestry: State Forestry. 1919/20 Forestry Department. 1920/21–1948/49 State Forest Service. 1949/50 Forest Service). Annual Reports, 1918/19–.
- N.Z. Royal Commission on Forestry. Report . . . Wellington, Government Printer, 1913 (A-J, 1913, C. 12).
- N.Z. Parliament. Parliamentary debates, 1854–.
- Stewart, W. D. Right Honourable Sir Francis H. D. Bell; his life and times. Wellington, Butterworths, 1937.

Thanks are due to various members of the N.Z. Forest Service for assistance especially with statistics, to Mr P. J. Lee and Miss B. Greenfield for detailed editing, to the late Dr A. H. McLintock for reading the text and to Miss N. M. Adams for the design of the dust cover.

A.L.P.

1

WHAT IS FORESTRY?

To DEFINE anything as diverse as forestry has always presented difficulties. Moreover, definitions have changed over the years with new thinking and new goals. An old definition—in vogue, say, a century ago—was that "forestry is the art of growing trees". Later, the application of scientific principles to this art gave a greater understanding of what the forester was doing when he grew and tended his stands of trees; the definition was then extended to "the art and science of growing trees". More recently there has come a period when, more than ever before, a great deal of what a forester does depends on the industries he supplies with raw material. In fact, with recent large extensions of artificial afforestation (planting of forests as planned and laid out by man) there has been a tendency to develop, more markedly than in the past, two distinct approaches to forestry: one from a dominant commercial motive, the other with the emphasis on growing and silviculture. A complete definition, therefore, embraces business; and forestry has come to be known as "the art, science and business of growing trees".

From time to time other ideas have been appended to the definition. For example, in many countries forestry has become more and more a public and national issue; and at times this has led to promoting the concept that forestry should, in the public interest, be a matter primarily of government policy. Such promotion has sometimes been carried too far. Without doubt the State moulds forest practices through its forest laws, but a State's laws are usually founded on national policies which reach far beyond the idea of forestry as a separate, isolated entity.

In a modern State three broad aspects of government policy affect forestry: general land use, foreign trade and the various implications of industrial policy (customs duties on wood, fostering of wood exports, maintenance of rural labour and unemployment relief). Each has its influence on forestry. Together they frequently determine the extent to which and the manner in which forestry is practised in any country. The forester must work within the framework of his country's domestic and international aims in forestry, which draws him into a political environment in addition to the forest one that is his immediate concern.

The incursion of the State into forestry can, in the long run, lead to disaster if forestry principles are departed from. It spelt the ruin of many forests in the pre-forestry days in Europe, for example, when States insisted on priorities for timbers for naval ships. These directives led to the devastation of forests in France and Britain in the 18th and 19th centuries; and the British Government's quest for spars for its navy even had a very small part in beginning the decimation of New Zealand kauri forests. During the Second World War the Nazi regime had to order the over-cutting of all German forests to get enough wood to carry on the war. There was some devastation, which would have become serious had the war lasted longer. This kind of thing will happen wherever the State allows utilisation demands to dictate where and when the axe shall be applied to the forest.

Forestry is a long-term venture but only those engaged in it from day to day are constantly aware of the vital time element. To ignore or discount it, leaves the forest resource vulnerable to switches in national thinking, development and policies. Although forestry practice must be flexible enough to be able to adjust to worthwhile changes, it must not be so susceptible to alteration as to be overthrown or harmed either by design or by the indirect results of self-interested actions of business and industry. Forestry must ensure that it is above all based on sound principles. Otherwise, any detrimental trends work insidiously against sound forestry over a long period. These sound principles are mainly concerned with silviculture and continuous supply (in the terminology of the forester "sustained yield") of forest produce. To ignore them spells the eventual ruin of any forest venture.

2

A FOREST POLICY EVOLVES

To ILLUSTRATE how far reaching forest policies can be, here are two examples, widely separated in time and space:

The first, quoted from B. F. Fernow's *History of Forestry*, refers to parts of Europe: "In the 12th and 13th centuries also, stricter order in the fellings and in forest use was insisted upon in many places. In the forest ordinances, which have always reference to limited localities, we find prescriptions like the following: The amount to be cut is to be limited to the exact needs of each family and the proper use of the wood is to be inspected; the timber is to be marked, must be cut in a given time and be removed at once; only dry wood is to be used for fuel and the place and time for gathering it is specially designated, similar to the present practice. The best oak and beech are to be preserved (this, however, with reference to the mast) and in the Alps we find already provisions to reserve larch and pine. The charcoal industry is favoured (because of easier transportation) but permitted only under special precautions. Bark peeling and burning for potash is forbidden. The pasture is regulated with regard to the young growth, and sheep and goats are excluded."

The second example concerns New Zealand and her afforestation policy. In 1960, the Labour Government approved a proposal to increase the total exotic forest area by one million acres by the year 2000 (an area slightly in excess of one million acres had already been planted), and by a further million acres by the year 2025. The proportions of these increases that it was hoped would be planted by private enterprise and local bodies and by the Forest Service were set out tentatively. This policy was confirmed by the National Government elected in 1961. It therefore had the backing of opposing parties in the House of Representatives. [The planting target has since been increased considerably.]

These examples emphasise the long-term nature of forestry, the need to plan its development, and the need to prescribe laws which, while directed at the forest owner, also involve the community in general; they also reveal the part that forests play in the use and protection of land and in the provision of forest products. Forests are of such importance in land use, and are normally such a permanent form of land use, that in forestry more than in most spheres it is necessary to adopt the principle of the greatest good for the greatest number of people, and to over-ride or control, where public needs demand it, the desires of the individual owner.

Sir Francis Bell (1851–1936)

New Zealand has been fortunate in having, at various stages in its development, statesmen who have seen the forest situation clearly and have laid, or endeavoured to lay, plans for the future —in other words, to formulate sound forest policies. Nevertheless, early attempts to do this failed, mostly because of the lack of administrative strength to put policies into operation or—and this was a matter of greater weight in the early days of settlement—through inability to win public and sometimes official support for long-term policies involving what was essentially conservation of native forest. Every effort in the young country was directed towards forest clearance and the development of what became remarkably successful agricultural pursuits. He was a brave man indeed who attempted to halt forest clearance anywhere. But by the early 20th century, destruction had become so rapid and the results had so often led to deterioration and erosion of the cleared land that one of the leading statesmen of the day, Sir Francis Bell, who had a deep understanding of and sympathy for forest problems, was able to achieve, by a few bold measures, outstanding success in laying the basis of a forest policy and in getting it implemented. Few people in any country have ever achieved similar success in laying the foundations of forest policies.

When speaking to the conservators of forests (officers in charge of State forests and forestry operations in a region) in the newly-created State Forest Service in 1921, Bell commented: "If we can keep a supply of timber for our children's children and their children's children, that will effect my aim. You must remember this, that the majority of the people of New Zealand, and perhaps the majority of the House, do not care anything about forestry. . . . We have got to be missionaries. We have got to show and prove that the principle of maintaining and establishing, controlling and managing our forests is a matter of public concern."

Bell, whose lifetime of public service included the offices of Crown Solicitor, Attorney General, Commissioner of Forests, and for two brief periods, Prime Minister, had taken a keen interest in forests and forest lands. These were then steadily diminishing and he was acutely aware of the need, despite the demand for farm land, for ensuring the setting aside of adequate forest lands. His legal knowledge showed him clearly that existing forest legislation (as incorporated in a State Forests Act of 1885) afforded no protection. His first move to bring about forest reform was to promote the separation of the office of Commissioner of Forests from that of Minister of Lands. This was done in 1918, and in November of that year Bell himself was appointed to the new office. Thus in 1919, when there was a wave of interest in forestry, stimulated by the shortages of timber brought about during the First World War, Bell was in a key position and the times were favourable for action.

Bell skilfully encompassed four matters that were all important to any forest policy.

1. He set up a new department of State with its own Commissioner of Forests, and so separated Government administration of forestry from the then highly political fields of land administration and forest clearance.

2. He piloted a Forests Act (1921-1922) through the House of Representatives, and this provided the legal basis for the sound administration of forestry.

3. Because of the nature of the forest reserves it created, the Act also ensured the retention of the remaining "protection" forests. (These are the forest-covered lands, particularly in the high country, which are essential to the wellbeing of much agricultural country at lower altitude.) The legislation that effected this is contained in the 1949 Forests Act, which repeats clauses of the 1921-22 Act concerning provisional and permanent State forest. These two categories broadly separate production and protection forest. The relevant clauses read:

"19. (1) The setting apart of any land as permanent State forest land shall not be revoked or altered except by Act of Parliament.

(2) Land set apart as provisional State forest land shall for all purposes of the Act be State forest land unless and until the Governor-General, acting on the joint recommendation of the Minister of Lands and of the Minister of Forests, shall, by further Proclamation, declare that the land is required for settlement purposes or for the purposes of a public reserve."

Preventing the release of permanent State forest except by Act of Parliament placed the onus of making a case for release firmly with the would-be clearer of forest. The new forest authority was put in a powerful position to resist forest clearance if it considered this to be unwise. But for this legislation, very serious and extensive inroads would undoubtedly have been made into protection forest. It also slowed up very considerably the clearing of indigenous production forest.

4. Bell also took the bold step of prohibiting the export of native timbers. The considerable overseas trade in these timbers was helping to deplete, much too rapidly, a valuable but dwindling resource. A few years later Government action to reinstate this export almost led to Bell's resignation from Cabinet, so strongly did he feel that the ever-increasing destruction of native forests had to be halted. In fact, by Bell's day the magnificent kauri forests of North Auckland, which had provided some of New Zealand's early national wealth, had almost disappeared.

Strong and effective action by Bell at the right time has enabled New Zealand forest administration and practice and the growth of forest industries to proceed from strength to strength.

Old-Established Forest Policies

Even in Bell's era the controlled use of national forest resources was not a new concept, though probably few New Zealand settlers were aware of this.

Many nations of Europe have for a long time not only retained a large part of their forests but also—and this is more important —managed them efficiently for production of timber, as protective cover for erodible land, and as places for public recreation. They have maintained their resources in the face of huge demands for forest produce and land for agriculture by large and growing populations.

The use of large areas for long-maturing crops, whether by the State or by private enterprise, must be the concern of every individual in any country, as all are affected by any damaging exploitation. For this reason many European countries have laws directed at preventing a reduction of the area under forest. A Swiss law specifically prescribes this; Austrian law makes provision for keeping the country's forest area at approximately its present size; and the laws of a number of other countries

A pa, or fortified village, of the natives in the province of New Plymouth (Taranaki), New Zealand. This sketch shows the lavish use of native woods by Maoris and the industry they applied to carving wood for their whares. (*Page 18*)

A tradition of building in wood grew up in New Zealand because first class native timbers were readily available and very cheap. Early settlers looking for house-building material, for firewood, for fencing, for a multiplicity of other purposes quickly discovered the virtues of native timbers. This is illustrated in the drawing of the beginnings of Nelson settlement.

(Page 20)

prescribe that any clear felling of forest shall be followed by reforestation. Because of the great increase in the demand for wood (a world-wide phenomenon) since the Second World War, some European countries have even set about increasing the area under forest in spite of the pressure on land for other purposes. Subsidies and tax remissions for new planting are common.

Keeping up a constant supply of wood to industries that depend on forests and maintaining the most efficient management of these forests for a sustained yield have become matters of national concern in most European countries. Swedish forest law contains a prescription relating these two things—sustained yield and good management—to the need for the forest owner to secure a satisfactory economic return. In Denmark, where great pressure on land developed long ago, forest policy as early as 1670 aimed at introducing the type of management into the forests that would give sustained production of timber. (The main concern at that time was for permanent supplies of firewood).

Where European forests help to protect the land on which they are growing this function is safeguarded strictly, even though wood is harvested from these forests. In protective forests that are privately owned working, felling and clearing of litter (the last a common practice in European forests) are rigidly controlled.

These are only a few examples of the wide scope of European forest laws and their effects on the individual and on the whole community.

Events of the Nineteenth Century

From the early days of European settlement in New Zealand (the 1840s) two developments influenced all moves towards formulating forest policy:

First and foremost was the rapid clearing of forest to provide land for agricultural development. The process of clearing yielded ample supplies of timber but did nothing to encourage any sort of management in the native forests. So long as forest was being cleared through the country there was plenty of timber, and the feeling was engendered that supplies were almost inexhaustible—still a common impression among millers of native forest. Some, however, saw that this clearance could not go on indefinitely and this conviction must have been driven home quite early through the clearing of steep country susceptible to erosion. The great danger of erosion would, in fact, be clearly appreciated by those who took the trouble to note the unruly nature of the rivers and streams before any forest was removed—a nature quite unlike that of the placid rivers and streams of England, from which many of the settlers had emigrated.

The second development was the early introduction of trees and the planting of these for shelter in the treeless, natural grasslands. These more or less experimental introductions were the basis of the highly successful exotic afforestation of later times.

The Seventies

Early in the 1870s provincial governments in Canterbury and Otago passed Forest Tree Planting Encouragement Acts, the principal provision of which arranged for free land grants in proportion to areas planted with forest trees on land already held. Provincial Governments, however, were abolished later in the seventies in favour of a central Government.

Julius Vogel, who as premier evinced an interest in forestry wider than the mere afforestation of treeless grasslands, introduced the first Forests Bill in 1874. He seems to have had more than one purpose in mind. Firstly some action was urgently needed to halt the rapid, and in many places indiscriminate, destruction of native forest. When introducing the Bill, therefore, Vogel eloquently and forcefully outlined the dangers of deforestation. He appreciated the importance of protection forest and the large part it played in New Zealand in protecting agriculture practised in the lower reaches of the rivers. He was also attracted by the commercial potential of forests, and included in the Bill a provision for setting aside areas of forest as security for the railway loans he wanted. But this provision was not acceptable to provincial representatives and had to be removed before the Bill was passed as the Forests Act (1874).

As a support for his Bill, Vogel tabled a report by a Captain Campbell Walker, then Deputy Conservator of Forests, Madras, on forest management in various countries. In 1876 Campbell Walker was appointed the first Conservator of Forests under the new Act. He spent a year in the country examining the forest situation and consulting people, preparing the way to set up a forest department. His fairly detailed description of forest conditions at the time is a most valuable record. However,

Campbell Walker was unpopular and so too was the Forests Act, which was being regarded partly as a legislative instrument to facilitate the abolition of provincial governments. Nor had the Act any support from ordinary settlers, who looked on forests purely as an obstruction to the development of farms. The Act was therefore "put on the shelf"; Campbell Walker presented his report to Parliament and left New Zealand in 1877. Public interest in forestry, as distinct from afforestation of the treeless Canterbury and Otago plains, waned until the 1880s. Enthusiasts among the landholders of the South Island continued to plant trees for shelter and for estate improvement.

The 1874 Act provided for an "annual sum of £10,000 for 30 years to be paid out of Consolidated Fund to the State Forest Account. . . .Money in the State Forest Account to be expended on forest management and development." It also provided "State forest to consist of lands as determined by General Assembly or of such provincial lands as set apart by Governor on request of Provincial Superintendent".

The tempo of forest clearance did not abate during the remainder of the 19th century. Timber was sometimes sawn as a byproduct of land-clearing for agriculture, but often it had to be destroyed without any attempt at use because there was no market for it nor any means of transporting it. The afforestation mantle of the Provincial Government of Canterbury was cast upon the shoulders of county councils and large grants of Crown land were made to Ashburton and Selwyn County Councils in Canterbury "for afforestation purposes". It now became apparent that the principal object of afforestation in those districts was shelter. The land granted to the counties was in long narrow strips, so that the whole countryside of the Mid-Canterbury Plains was cut up in a large draughtboard pattern; and from that date to the present day, the principal counties there have pursued an active policy of planting and maintaining these areas.

Not all of them are yet planted—and indeed the chances are that many of the vested areas will now never be planted; as an example, the Selwyn Plantation Board in 1949 reported that it had planted 13,000 acres out of a total of 18,000 acres. At first sight, this result seems unsatisfactory, over a period of 60 years after the land vesting grants; but its history is in fact reasonable, and gives a good example of the manner in which all forestry proposals, no matter how carefully designed, are modified by changing economies and changing national philosophies.

The original land grants in the era of abundant land were laid out with due regard to the directions of prevailing winds and to a reasonable distribution for shelter purposes over the whole countryside. It would seem that no consideration was given to land quality, unless possibly by an avoidance of land lying in depressions in which trees would not afford shelter to adjoining areas. The whole scheme was indeed a miniature "shelter belt" programme, anticipating by half a century the American project, on a continental scale, which made the term familiar to forest administrators the world over. At that early date, subsidies and subventions in money, in labour, and in kind (free trees on a munificent scale) from central Government sources were not the commonplace of a national administration that they were to become later in many parts of the world. The Government had provided free land generously; the recipient counties and their ratepayers were expected to provide the trees and the wherewithal to plant and protect them. Plainly the afforestation of thousands of scattered acres could not be organised and executed in a year or ten—even supposing the money were available from normal county revenues, which it was not. The device was suggested, and approved, that planting should be gradual and that the portions on which planting was deferred might be leased to provide revenue that would defray at least some of the immediate costs. This was done, and continuous smallscale planting so financed became a routine accepted by all.

This brief description has taken us far beyond the immediate subject of forestry of the nineteenth century. It seemed desirable to trace rapidly its full growth at this stage, because probably no other single factor has so largely influenced the New Zealand public's conception of forestry. In that view, forestry is—or was until recently—afforestation of bare land followed by clear felling and replanting. The idea has been responsible for the ready provision of many millions of pounds of money for "forestry", a result which frequently has made the New Zealand forester the envy of his overseas colleagues. On the other hand, it has often, again until more recent times, been responsible for the refusal of moneys and, what is so often worse, the refusal of authority for the forester to deal with any question which does not entail planting exotic forest species on bare land. It was responsible for segregating afforestation, the disposal of native timber, and the administration of reserves, into more or less distinct compartments in the Lands Department, a development that will be described later. This is the other side of the picture which, in the past, sometimes dismayed the New Zealand forester.

Although the Forests Act 1874 was shelved for a time forestry was far from being a dead issue, attention being drawn to it by rapid forest clearance and the needless destruction of a great volume of valuable timber. In 1879 a retired French forester, A. Lecoy, read a paper to the Wellington Philosophical Society entitled "The Forest Question in New Zealand". (Trans. N.Z. Institute Vol. XII 1879). His opening paragraph is worth repeating: "Parallel with the Public Works system might be initiated a new policy, tending to promote the interest of the Colonial Treasury, by improving and consolidating, instead of exhausting, the revenue derivable from the public estate, by a systematic treatment of the Crown forest lands, which revenue might be increased to such an amount as to provide at any time for the largest portion of the expenditure required for general State purposes. Had measures in that direction been carried out ten years ago, when in the House of Representatives Mr Potts moved 'That it is desirable Government should take steps to ascertain the present condition of the forests of the colony, with a view to their better conservation'—had the forest question been then more practically investigated and considered in all its aspects, especially in that of the income which State forests, under systematic treatment, can afford to the public purse, without either their climatic advantages being disturbed through the fellings, or the supply being reduced below the demand, as evidently would have been (and still would be) the case in New Zealand—we might have had by this time the same extent of railways, a less heavy indebtedness, and also fewer alienations of valuable timber lands. Furthermore, we should have now a surplus income, which fund would afford a means for a more equitable distribution of the public revenue towards municipal interests than the present allotment of lands for such purposes can allow."

Lecoy also submitted suggestions to the Minister of Lands on the proper management of New Zealand forests, the need to regulate cutting, and the need to provide for regeneration to ensure continuous yield.

The Eighties

In 1882-83 conservation of forest was the subject of two motions in the Legislative Council and of a question in the House. The motions, submitted by the Hon. Mr Chamberlin, referred to Lecoy's report to the Minister of Lands. Chamberlin said that he was induced to give notice of it by the information he had been afforded by a return he had asked for of the forest reserves of the colony. He was sorry to say that, looking over that return, it appeared to be very unsatisfactory, for the total area reserved for forest conservation in the colony was only 581,000 acres, the colony itself containing 64,000,000 acres.

The question in the House quoted a long section of the 1877 Campbell Walker report. The question, put by Mr Cadman, asked the Government when they intended to take any action towards supplying, "the want, shortly to be felt in this Colony, which was now being caused by the wholesale denudation of our forest lands"?

Vogel had maintained his interest in forestry and was probably responsible for getting Thomas Kirk, Professor of Natural Sciences, first at Victoria University College, Wellington, and then at Canterbury Agricultural College, Lincoln, to tour the forests. In 1885 Vogel introduced the second State Forest Bill, which was read a second time without acrimony and passed by the Legislative Council without debate. It gave authority for the setting aside of State forests; for the establishment of a school of forestry; and for the appointing of a forest staff.

Some impulse for the passing of the Act had come from North Auckland and from those interested in kauri forests, because the statute specifically provided that the school of forestry should be at Whangarei, and set aside an area of some hundreds of acres merely for the school. In addition, over 19,000 acres of mainly kauri forest at Puhipuhi, just north of Whangarei, were proclaimed State Forest. No doubt the Government had in mind the permanent management of kauri forest.

In 1886 Thomas Kirk was appointed Chief Conservator of Forests, under the Act of the previous year, and he made plans for the forestry school and for the development of forestry. However, the juggernaut of politics and economics apparently quickly extinguished the brief spark of interest in forestry, the services of Kirk and most of his small staff being terminated. It is customary to deplore these setbacks and to speculate wistfully on what might otherwise have happened, but it is by no means certain that much could have been expected to develop. The country, the people, even the foresters, were not prepared or able to pursue an active policy in forests, the structure and composition of which are even yet little understood.

A few State forests were created under the 1885 Act. They

were useful, but persistence in administering the Act might have created many more. However, their existence would certainly have impeded the rapid and successful development of land for cropping and livestock raising, and forestry would have collected around itself such a legacy of antagonism that it is at least conceivable that more harm than good could have been done. No human being in the mid-eighties could have forecast the possible boundaries of successful agricultural development of the forest land then yet untouched. No science or scientific knowledge of that time (and by modern standards there was little of relevance) could have been of assistance. The times were not ripe for forestry of primeval forests, except that forestry policy of complete protection, which has always aroused bitter animosity in democratic countries.

Early Twentieth Century

Although the second Forests Act was set aside in a similar manner to the first (see pages 8-9) and administration of forests remained under the various Lands Acts, interest in forests and forestry continued to grow. A Royal Commission on Forestry in 1913 defined some of the main forestry and timber problems of the day. The commission's secretary was the celebrated Dr Leonard Cockayne, a plant ecologist of world renown and one of a handful of men who laid the foundation of this science. He studied and wrote a great deal about New Zealand forests. The First World War prevented any action being taken on this commission's recommendations, which were shelved until Bell revived the forestry issue.

At the outbreak of the 1914-18 war Britain was the largest importer of wood in the world, huge quantities being bought from Scandinavian countries and smaller amounts from Canada. During the war Britain was cut off from the most important sources of this supply, and on all ships food and munitions naturally had priority for the limited cargo space. Nevertheless, wood was an essential commodity for war. Britain had to plunder her own meagre forest resources—at that time she was the most poorly forested country in Europe and most of the woodlands and forests were in a deplorable condition. Such dependence on a very small and poor forest estate to supply an essential commodity in a time of crisis focused attention on the need to develop a long-term forestry policy in Britain to avoid a repetition of the situation. As a result forestry and forest products received a great deal of attention after the war, and Britain soon began a very large forest planting programme.

This interest spread through the British Empire; it was common in those days to think and speak of a coming wood scarcity. The theme fitted readily into the New Zealand scene, where awareness of the importance of timber had been heightened by shortages during the war—caused, however, by lack of sawmill capacity and manpower and not by lack of resources. The cry of a coming wood famine was widely repeated. Bell, therefore, found public and political support for the measures he wanted to introduce, and before long (1920) a vigorous personality, L. MacIntosh Ellis (a Canadian forestry graduate, who had served in the Canadian Forestry Corps in France) was appointed Director of Forestry of a new State Forest Service. Before this, forestry activities had been within the Department of Lands segregated in three sections:

(i) State afforestation, using exotic forest trees, had been begun at the end of the previous century. The need for it was generally admitted and forestry conveyed to most people the meaning of afforestation. It had the background and benefit of fifty years or more of tree introduction by settlers and of experimental planting, mainly for shelter, in the low-rainfall treeless areas. It had begun formally as a separate branch of the Department of Lands under a Chief Forester, who was appointed in 1896. The first central state nursery was established at Eweburn on the Maniototo plain in Central Otago; not by accident, but rather because of the thinking of the day, was this treeless area chosen for the first State afforestation.

(ii) The selling of native timber from Crown land was the responsibility of another branch of the department. The timber came from land required for settlement and from the few State forests that had been proclaimed under the 1885 Forests Act, under which land could be transferred to and from State forest status as occasion seemed to demand.

(iii) The growth of a public feeling for forest reserves, a reaction to the ruthless clearing of native forest, found expression in the promotion of scenic reserves, national parks, and public domains, administered by the Department of Lands under a special series of statutes and regulations. An Inspector of Scenic Reserves was appointed to deal with what was essentially a part of forest administration.

The segregation of these three forestry activities of the Department of Lands continued for almost 20 years into the 20th century. One effect was that forestry continued to be interpreted as afforestation; the zeal displayed and progress achieved by those entrusted with its development seemed only to strengthen the idea that they were men apart, men entrusted with work of such a special and peculiar nature that they could have neither time nor aptitude for the difficult work of indigenous-forest management and indigenous-timber sales.

Early in the century a succession of untoward events had prevented any such effective administrative advance as had been visualised when the forest Acts were passed. The death of Seddon, the Prime Minister, in 1906 shook the stability of the Government that had encouraged the Afforestation Branch of the Department of Lands 10 years earlier; and in 1909 the Chief Forester died. These events and then the general disruption caused by the First World War prevented administrative advance in what was still a minor and little appreciated activity.

Nevertheless when the war ended and Britain began to urge on her dependencies and colonies the importance of forestry as an essential Imperial development, New Zealand did not have to start from scratch or from behind scratch. She had at least one phase—afforestation—well established and an experienced working cadre. But even the enthusiasm of unusually well-informed forest officers cannot work up afforestation quickly from nothing to a peak.

The new State Forest Service was at first controlled by an officer who had for some years been Inspector of Scenic Reserves for the Lands Department and an enthusiastic conservationist. This appointment plainly signalled a new national philosophy. It united under one control the exotic forestry of the afforestation branch, the hitherto unattempted indigenous forestry, and indigenous forest regeneration. Politically, the move was in the care of Sir Francis Bell, who became the first Commissioner of Forests and who had the advancement of forest policy and forest practice very much at heart.

For the first time, publicity was enlisted to support the cause of forestry. Sir David Hutchins, one of the earliest foresters in the British Colonial Service, who was living in retirement in New Zealand, was encouraged to prepare reports on forest potentialities. Sir David had spent a lifetime organising and publicising forest management in Cyprus, South Africa, and Western Australia. His vigour and zeal needed little stimulus and he was prepared to preach forestry in and out of season. He wrote *New Zealand Forestry, Part I. Kauri Forests and Forests of the North and Forest Management, 1919*. With the whole Empire, led by Great Britain, advocating and re-organising forestry in the light of wartime scarcity of timber and shipping, there was little difficulty in finding audiences for all forestry propaganda; and the firmly established nucleus of the afforestation staff gave New Zealand a good start in the Empire-wide campaign of forest overhaul.

Fortunately, the new Forest Service was too well led and too farseeing to rest at a mere acceleration of afforestation. It appreciated the opportunity offered for, and the need for, dealing with the existing timber trade and timber titles of indigenous forests, as well as for growing wood for future overseas markets. Bell recognised the urgency for a revision of an obsolete and ineffective and inoperative State Forests Act of 1908—a rehash of the 1885 Act.

Forests Act 1921-22

A new State Forests Act passed in 1921-1922 provided for adequate administration. This time there was no faltering as there had been following the Acts of 1874, 1885, and 1908. Recruiting began for the staff the Act prescribed, the nucleus of which was available from the former Forestry Branch of the Department of Lands and Survey.

In a report* to the Commissioner of Forests in 1920 the Director of Forestry had included the following among his recommendations:

A Forest Development Fund for forest development and demarcation.

The administration and management of all the Crown forests and forest lands by the Forest Service.

A progressive timber-sale policy.

Adequate facilities for technical education in New Zealand.

State cooperation in private tree-growing by various means such as equitable taxation, forest fire insurance, and forest fire protection.

Forest Service administration and management of all scenic reserves, national parks, forest reserves, forested national and educational endowments, and forested native lands.

* *Forest Conditions in New Zealand and the Proposals for a New Zealand Forest Policy*. L. MacIntosh Ellis.

A Forest Products Laboratory and Bureau of Forest Research; survey and inventory of the forests, forest resources, and soils of the Dominion.

An economic survey of the timber industry and of the timber-using industries.

Administration and protection of fish, bird, and game resources by the Forest Service.

The introductory section of the director's report had included the following:

"The writer hopes, for the sake of the continued prosperity and welfare of New Zealand, that his attempt to bring about a better state of affairs as expressed in this report will receive more than a hard-boiled and indifferent interest, and that the proposed remedial legislation will reach the statute-book in a sound, virile, and workable form."

3

NATIVE FORESTS AND THE TRADITION OF WOOD USE

THE POLYNESIAN immigrants to New Zealand travelled the oceans in canoes fashioned from wood. Once settled here they quickly discovered tree boles—particularly of kauri (*Agathis australis*) and totara (*Podocarpus totara*) easy to hollow out with primitive tools. They must have prized the size, regularity of shape, and lack of taper of the boles of these trees and the durability of the timbers. Rimu (*Dacrydium cupressinum*) also was sometimes used for canoe making, but was a heavier timber and not so easily worked.

Trees from which large canoes could be made became so important that possession of them sometimes led to tribal warfare. A chief would often reserve a tree for a young son by placing a *tapu** on it to ensure that it would be there when the son grew up—a form of conservation by a primitive people. The same timbers were used for buildings, and totara was the main timber for carvings. It is interesting to speculate to what extent the availability of such an excellent carving timber led to the very high development of this Maori art.

Wood was also used by the Maoris to provide tools for working the soil. Cook described their main implement of cultivation, the *ko*, as a "stout picket"; this was simply a sharp-pointed stick, usually made out of manuka (*Leptospermum spp.*) or maire (*Gymnelaea spp.*). A *hoto*, or wooden shovel, was also used, as were various paddle-shaped tools, for planting and cultivation.

The Maori had a few primitive musical instruments all fashioned from wood, usually matai (*Podocarpus spicatus*). He also had a war gong, or *pahu*, which Johannes C. Andersen has described as "a slab of totara or matai, sometimes thirty feet in length, two or three feet in breadth, and six inches in thickness, suspended on cords some height from the ground. It was used for signalling, and when struck with a heavy beater of maire,

---

* Ban on any interference.

the resounding boom resulting might, under favourable circumstances, be heard to a distance of from six to ten miles". (*Maori Musical Instruments*. Art In New Zealand 1929. II. 5. p. 91).

The Qualities of Kauri

About the time Napoleon's armies were sweeping over Europe and Nelson was gaining a supremacy over the French navy, traders were beginning to make use of kauri. Before that, explorers had discovered the tree. It has been recorded that Cook in his voyages to New Zealand did not find the kauri, though this seems almost inconceivable, and further research may show that he did. But almost at the same time as his voyages, the French of Marion du Fresne's expedition in 1772 brought down what is believed to have been the first kauri tree felled by Europeans. It was to have been used to replace a damaged bowsprit, but had to be left where it was felled when the expedition was attacked by Maoris.

England in the first two decades of the 19th century was searching for timbers to build naval ships. English oak was used for framing and waterline planking, Scandinavian pine for planking above waterline and for spars and, before the War of Independence, white pine (*Pinus strobus*) from the American colonies. Navy needs had helped to stimulate British penetration into the American colonies. The need for good timbers, especially for masts, was sufficiently great for the Admiralty to take note of the rumours of their presence on the other side of the world in New Zealand. No doubt by this time some samples of kauri had reached England. In 1783, a James Matra proposed to the Government that: "It may also be proper to attend to the possibility of procuring from New Zealand any quantity of masts and ship timber for the use of our fleets in India, as the distance between the two countries is not greater than between Great Britain and America. It (kauri) grows close to the water's edge, is of size and quality superior to any hitherto known, and may be obtained without difficulty."

The first real trade in kauri from New Zealand was pioneered by merchants from the growing Sydney settlement; and in 1840 Thomas Laslett, Timber Inspector for the Admiralty, visited New Zealand to inspect suitable trees for masts. It has been said that some of the British men-o'-war in the Battle of Trafalgar (1805) carried masts of kauri, but this is most unlikely. The export trade in sawn timber—at first pit sawn—to

Australia grew rapidly and expanded to other parts of the world. The milling of kauri did not cease until the great kauri forests of North Auckland and of the Coromandel Peninsula, thought to have covered about three million acres, had all but disappeared. Whatever regrets might be felt about this, the use of kauri for every conceivable purpose to which wood could be put, and its export overseas, built up skills in conversion of wood and in its use for housebuilding, shipbuilding, and many other purposes.

A tradition of building in wood grew up in New Zealand because first-class native timbers were readily available and very cheap. Early settlers looking for timber for house building, for firewood, for fencing, and for a multiplicity of other purposes quickly discovered the virtues of the native timbers (principally, kauri, totara, rimu, kahikatea (*Podocarpus dacrydioides*), matai, silver pine (*Dacrydium colensoi*), and tanekaha (*Phyllocladus trichomanoides*).

The superlative qualities of kauri had been proved by whalers and traders and by the British Royal Navy even before organised settlement began. In fact it was the use of kauri that did most to lay the foundation of wood use. Kirk in his *Forest Flora of New Zealand* (1889) wrote the following long but well deserved testimony to kauri wood:

"Kauri is unquestionably the best timber in the colony for general building purposes, ground-plates, beams, framing, rafters, joists, flooring, and weatherboards; also for roof-work, dadoing, panelling, mouldings, sashes, doors, and all kinds of joiners' work, as well as for decorative fittings, whether in public or private buildings. It is largely used for railway-sleepers, bridges, wharves, and constructive works generally; for telegraph-posts, mine-posts, mine-props, caps, struts, etc.; for the masts and deck-planking of ships, for which it is unsurpassed, being regular in the grain, free from large knots, of smooth even surface, and able to resist a large amount of wear. It is also largely used for the outer and inner planking of coasting craft and boats.

"It affords the best timber for seats in churches and other public buildings, as it takes a high polish: it is of equal value for bank counters and fittings. Kauri is adapted to all the purposes of the cabinetmaker where a light-coloured wood is required: ordinary wood is excellent for common furniture, or for framing as supports for veneer: figured and mottled varieties are highly prized for ornamental work. It is highly valued for coopers' ware.

"It is largely used for fencing-posts and rails, palings, and shingles, both sawn and split, and for tramway-rails.

"A large quantity of second-class timber is utilised for packing-cases, tallowcasks, shedding, and other temporary purposes.

"It may safely be stated that no other New Zealand timber is capable of being applied to such varied uses."

The tradition of wood use has pervaded all New Zealand forest policy since the early days of settlement. Those who recognised that at some time supplies of native timbers must reach a very low level were not concerned to replace wood with substitutes, but to renew supplies through afforestation.

Wood Use Becomes Government Policy

It was the realisation that forests and wood were essential to the community that caused Sir Francis Bell to introduce the forest policies for which he had much public sympathy, but against parts of which he also had the opposition of the strongly entrenched timber trade. To ban the export of native timbers, as he did, was a severe blow to an established trade, but an essential measure in the conservation of native forest and timbers. It was the first such step since the founding of the colony and paved the way for later measures. His concept of creating permanent State forests was based on the same acceptance of wood as a primary raw material. He was, without doubt, thinking that native forests would yield to management, for his own words were: "The forestry I want to initiate consists, first and foremost, of conservation and use of existing forests, and secondly, and far behind, plantation."

Even so, it took 10 years from the imposition of the ban on exports of native timbers, which were running at the rate of over 70 million bd.ft per year to run down to about half that quantity, and not until 1950 did the quantities become negligible. Over the past few years limited export has again been allowed.

Afforestation to Perpetuate Wood Supplies

To perpetuate supplies of wood the Government began afforestation almost at the close of the 19th century. Planting proceeded at a modest rate, so that by 1920 between 30,000 and 40,000 acres had been established. The creation of the State

Forest Service, combined with a set of circumstances after the First World War, led to an afforestation boom in which both State and private enterprise participated. This reflected an appreciation both by the Forest Service and by Government of the depletion of the indigenous timber resources, a depletion that became particularly apparent as a result of a thorough survey of the remaining forests. This survey was conducted without the aids of aerial photography and statistical sampling, which were available for a resurvey carried out more recently. Nevertheless, its results proved, with the passing of the years, to have been surprisingly accurate.

The expanded planting lasted from about 1925, in which year 17,000 acres were planted, to 1936, when 30,000 acres were established. The highest acreage planted in any one year was just under 100,000 acres, and the total was 720,000 acres or about half of our present total planted area. Most was of radiata pine. The hot pace for a young country led to many departures from forestry principles. However, the quantity of wood produced 25 to 30 years later has dominated all events in the forest development of the country and determined much that is happening now and that will happen in the future.

Change from Indigenous to Exotic Timbers

Some of the major effects of large-scale afforestation are dealt with later but of relevance to the theme of a tradition of wood use is the fact that the nation was presented quite suddenly in terms of forestry time with a very large volume of poorly and rapidly grown wood, the greater part of it radiata pine, as replacement for high-quality native timbers which had been used so successfully.

The vigour with which problems associated with the use of radiata pine have been attacked is a credit to all involved and a reflection of the knowledge of wood and its use possessed by many people in the country. In less than twenty years from the time large-scale sawmilling of exotic timbers began, their sawn timber output exceeded that from native timbers. Moreover, the exotic timbers have established themselves firmly in many uses in preference to native timbers, and, as better grades become available, because of improved growing, and properties become more widely known, their uses have become manifold. Of even greater importance, and with the promise of large and profitable developments in future, has been the establishment and rapid

Pitsawing of kauri logs. A trade in pitsawn kauri timber sprang up between North Auckland and Australia even before the days of organised European settlement in New Zealand.

(Page 20)

The position that wood came to occupy as the traditional building material and the skills that developed in the use of it, are amply illustrated in the Government Buildings, Wellington, built in 1876. Timbers used were: weatherboarding, kauri; flooring, matai; joists, rimu; and framework in first section—jarrah. Statistics: Length, 255 feet; depth, 161 feet; floor area, 101,300 sq.ft. The total amount of timber used including alterations and subdivision was approximately 1,100,000 super feet.

growth of pulp, paper and board industry based entirely on exotic wood, mainly radiata pine. This industry has revealed further virtues of the wood.

In the long run, and barring calamities, the economic effects of introducing radiata pine into this country must be very great indeed. Its success is now a primary consideration in New Zealand forest policy. At present radiata pine dominates the forestry scene and will do so for some time as far as exotic forestry is concerned; other introduced forest trees have been very successful but are taking longer, usually because of slower growth, to find their spheres in afforestation and utilisation. The stage has been reached, however, when private enterprise plants principally radiata pine, while the State has stabilised on three main species, radiata and Corsican (*Pinus nigra*) pines and Douglas fir (*Pseudotsuga menziesii*). Extensive trials are being undertaken of varieties of these trees and of many other trees, so that the selection of species or varieties for afforestation is far from final and can be expected to be changed by the addition of further species as knowledge of them becomes available. Radiata pine itself is being subjected to detailed analysis and research.

It is essential, principally for insurance against disease or pest attacks on any one species, to grow several, but it is undesirable to grow many. In general, the simpler and more uniform the raw material supplied to industry, the more efficient the industry becomes and thus the more profitable it will be. This applies particularly to industries using large quantities of wood. Industries such as furniture and cabinet-making use a variety of timbers in small amounts, but their needs can be catered for by developing the use of secondary native timber trees and by imports. The forester must strike a balance between his desire and the need to grow a variety of trees and the need to supply industries with what serves their purposes best. He must never be guilty of growing unthrifty forests by planting trees required by industry on sites that will not grow them well.

It is doubtful if any other country at any time has experienced such a revolution in wood use as has occurred in New Zealand with the move from indigenous to predominantly exotic wood. However, even before the substitution of exotic for native timbers there had been changes in uses of the latter. While kauri was freely available and cheap it was the preferred building timber. Most of the houses throughout the north and many elsewhere were constructed of kauri, and large consignments were distributed throughout New Zealand. The four-storey Government building completed in Wellington in 1877 after three years' work was largely kauri. This very large wooden building, still in use, is likely to survive for a long time yet.

As kauri became scarcer and rimu came to be better known as a timber, so building turned to it. Rimu soon became the main construction, joinery and furniture timber and, somewhat later, the main veneer wood. The change from kauri to rimu was merely from one high-quality wood to another. However, rimu contained many more defects and could not yield the clear widths and long lengths which had been readily cut from the great cylindrical logs of kauri. Rimu also contained proportionately less heartwood and was, therefore, in general terms less durable. It was a much wetter timber than kauri and therefore not so quickly usable. Nevertheless, rimu was a high-quality timber and the top grades were very free of defects and warping troubles and well suited for finishing timbers. Its distinctive figuring extended its use in furniture manufacture to decorative work.

Other native softwood timbers used for more specialised purposes were of even higher quality in certain respects. Totara was one of the choice joinery timbers and was also employed where durability was important. Matai was used for joinery and high-quality flooring; for a softwood it is exceptionally hard.

Rimu remained the principal timber cut until about 1950. After that it was overtaken by radiata pine, which now has an annual production considerably greater than the total cut of all other softwoods. It began to secure its position during the Second World War with such success that it took only about 10 years to supersede rimu as the main timber for house framing, and about 20 years to dominate the timber market completely. That such a change could have been so effective was made possible only by careful planning, bold promotion and, especially, research.

Production by species for the years ending 31 March 1947 and 1967 was as follows:

Production of Rough Sawn Timber for Years Ended 31 March 1947 and 1967

Indigenous	1947	1967
	(million bd.ft)	
Rimu and Miro	172.1	165.5
Matai	19.7	22.3
Totara	9.6	8.6
Kahikatea	15.1	16.9

	1947	1967
Kauri	2.3	1.1
Tawa	7.6	17.0
Beech	11.6	13.4
Minor species	1.7	2.6
Total	239.7	247.4
Exotic		
Exotic pines	111.6	465.3
Douglas fir	—	27.1
Eucalypt	1.2	1.6
Minor species	1.5	6.2
Total	114.3	500.2
Grand total	354.0	747.6

The State as Pioneer

Radiata pine was non-durable. Most of the pine from millable stands hastily established in the boom planting years had many defects, especially by comparison with the high quality native timbers which had been freely available for so long. Industry was certainly not at all anxious either to saw or to accept radiata, nor customers to buy it, even though a fair amount of experience in its use had been gained in Canterbury, and it had been widely used for fruit cases and crates. The State Forest Service itself, therefore, began logging and sawmilling in Rotorua, the Waipa State Mill being built to saw exotic logs, principally radiata pine. It introduced to New Zealand the Swedish gang-frame saw. The object was to develop both production and marketing techniques for the sale and use of exotic timbers.

For possibly a decade the Forest Service was the largest logger and sawmiller of exotic timber. During that time methods of sawing and grading radiata pine were devised; grading studies were accompanied by extensive testing of the physical and mechanical properties of the wood; seasoning practices were investigated, both on laboratory and commercial scales; and preservation, the most important process to master in dealing with non-durable timber, was also worked out on laboratory and commercial scales with, fortunately, outstanding success. Logs of radiata pine from trees thirty to thirty-five years of age—an average rotation for New Zealand—are virtually all sapwood, which is readily penetrable by all timber preservatives in commercial use. Here, then, was a way of producing a timber even more durable than any that had so far been available to the building industry, even from the fine range of native woods.

The investigations and commercial operations undertaken by the State paved the way for industries to grow up and to produce, treat, market and use radiata pine satisfactorily. Waipa State Sawmill is one of the most successful and powerful tools of forest policy in New Zealand.

With these new industries there has developed a more scientific approach to the use of timber. Quality controls have been insisted on by institutions lending money for house and other building construction, because their early standards were based on the good native timbers. A Timber Preservation Authority was set up by statute in 1955 to promote and safeguard standards of preservation. Grading of timbers is constantly under review by the Standards organisation.

Meanwhile attention was being turned to the use of radiata pine for production of pulp and paper. A lead to the possibilities of this has been given by a very modest amount of research in the 1920s, and the starting of paperboard making by the Whakatane Board Mills Ltd. in 1939. At first this firm used eleven-year-old radiata pine thinnings for some of its raw material. Some years after the war the pulp and paper industry began in earnest and grew rapidly, a feature being the integration of manufacturing units with sawmills to ensure maximum use of raw material. The development of the large commercial enterprises is discussed on pages 81–85.

Recently, advantage has been taken of the ease with which the round produce of radiata pine can be treated with preservatives to use natural rounds as posts and poles. Such treated produce from two other exotic trees, Douglas fir and European larch, had earlier found a market. Radiata pine, because of its availability and ease of handling, quickly took over a large part of the post and pole market.

What began with the building of some of the first European dwellings in a superlative timber, kauri, has developed, under the guidance of deliberate policies, into thriving forest products industries. The stage has been reached when wood from artificially grown forests of exotic trees should produce most of the domestic requirements of timber, round produce, and pulp and paper. So successful, indeed, has been the growing of exotic forest trees that the State now looks to forest products to supply commodities for export and has adopted a policy of encouraging the growing of adequate raw material for this purpose.

Place of Radiata Pine

Radiata pine dominates the production forestry scene, and promises to go from strength to strength as long as there is suitable land to grow it on. The foundation to this dominance has been laid with produce from untended forests. The produce from tended forests is far better and much of the forest recently planted is being tended.

Other exotic forest trees are important, despite their smaller areas and the long period they take to mature. Most promising is Douglas fir, which can produce a strong structural timber in large sizes when trees are old enough to yield big logs. Timber from the virgin stands in its native home, the west coast of North America, has gained a world-wide reputation and virtually a world-wide market. Some of it is still imported by New Zealand. Though trees grown in this country do not produce timber as good as that from virgin natural stands, the basic characters are the same, and all the timber that planted stands will yield is being used. Increasing quantities of timber from pines other than radiata are being marketed, but use will be determined very largely by the pattern set by radiata.

Although such good progress has been made with the growing, harvesting and conversion of exotic forest trees, only the most naive optimist would maintain that it is now safe to "liquidate" indigenous timbers as fast as industry wants them. Present exotic forests are highly vulnerable to fire, consisting as they do of large areas of one species and often of one age class. For the same reasons they are vulnerable to insect pests. They are still infants, although lusty, and it would be unwise to regard them as anything else.

Conservation of Indigenous Timbers

For these reasons governments maintain a policy of conservation of indigenous timbers. This policy, introduced by Bell (see page 6) was originally directed at conserving native timbers while a sufficient exotic forest estate was being grown. This has come about, but conservation is still needed and takes the form of spreading the release of State Forest indigenous softwood resources for some time into the future. As no control has been imposed on the milling of privately-owned native forest (including Maori-owned forest) State resources must bear the total burden of the policy. Whatever the effects of conservation and however optimistically we may regard the results of indigenous forest management, New Zealand must, in the not far distant future, look to the exotic forests for nearly all supplies of forest produce. Indigenous forests, because of their much slower growth and the fact that most of the timber available for milling in the distant future will be hardwood (southern beech), can be expected to yield only a small part of total requirements.

4

FORESTS AND LAND USE

SWISS FEDERAL forest law contains the imperative and, to New Zealand ears, startling clause: "The forest estate shall not be reduced." Behind this simple law lie centuries of experience in mountainous country where all land use must be practised with the utmost care. Above all other things, forests were recognised as being essential for protective purposes: to keep soil on mountain slopes, to help regulate mountain torrents and stream flow, and, as mountain villages developed, to protect them from avalanches. Then, as population increased, wood became an essential raw material. Further growth of population led to heavier demands for food, and to produce it more and more forest had to be cleared. In time the State was forced to decide what forest was essential for protective purposes and how much should be maintained to give the harvest of wood which the community, including industry, had become accustomed to receiving and using. At the appropriate stage the State took the simple action of prescribing that "the forest estate shall not be reduced", instead of introducing an elaborate system of controls.

Other European countries have had to face the same problem, though because of differing geographical patterns and histories they have solved it in different ways; but all have placed emphasis on the retention of an adequate forest estate in spite of growing populations requiring more and more food.

The lesson for younger countries is that forests are an essential part of land use and that they have important roles as civilisations develop. The pattern is determined by influences from various directions, but first and foremost is the need for forests to conserve soil and water and to yield raw material. In the light of these experiences we can examine changes that have occurred, and are still taking place, in New Zealand forests.

Indigenous Forest

There is growing evidence that fires during the Maori occupation of New Zealand (which began about A.D. 1350) destroyed large areas of forest, particularly in the north and east of both the North and South Islands, where conditions can become very dry. Geologically recent eruptions in the centre of the North Island also destroyed some forest, though possibly piecemeal. Natural tussock grasslands occupied some sixteen or more million acres, or almost a quarter of the country, and mountain land above the timberline—4,500 ft in the north to 2,500 ft in the south—was clothed with low-growing mountain vegetation. At the time European settlement began in earnest (1840s) the remainder of the country was covered in forest. Those who have studied the evidence consider that this amounted to about two-thirds of the total land area.

Clearance of Indigenous Forest

Settlement of the land proceeded most rapidly at first in the open tussock grassland because of the ease with which it could be cleared by burning, worked with implements, or stocked immediately with sheep. But the generally good fertility of soils from which the mixed native forest had been cleared soon became apparent. The clearing of forest, which the Maori had initiated for his shifting cultivation, was therefore greatly accelerated.

Of outstanding importance in exerting some control in this clearance (though probably not fully realised at the time) was that after the signing of the Treaty of Waitangi in 1840 Maori land could be disposed of only to the Crown. A semblance at least of law and order was therefore introduced into land settlement. Of greater importance was Government control of unsettled Crown land, which was essential should there be a need to introduce major policy changes in land settlement. (This has, indeed, been found desirable on a number of occasions.) The flexibility thus provided has been used to set aside State forests, and it allowed Bell to create his provisional and permanent State forests. Government was also able to reserve the forest, absolutely essential for protective purposes, along the main mountain chains of New Zealand. Even so, in many places the clearance of protection forest has gone beyond the bounds of safety. Without the safeguard of Crown ownership, however, a great deal more could have disappeared under the pressure on land for livestock grazing and the results would have been disastrous to the lower reaches of many rivers.

In the coniferous forests (pine, spruce, etc.) which originally covered a great part of the temperate Northern Hemisphere,

Waipa State mill and post and pole yard. Whakarewarewa forest. This plant was started in 1939 to demonstrate the modern milling and marketing of exotic logs. It is now a substantial business, although it still serves many demonstrational purposes and has proved to be a powerful tool of forest policy.

(Page 26)

Wanganui River valley, Westland. This is typical of the important protection forest that remains. Were it to be removed, aggradation of the riverbed would proceed at a very rapid rate. About two-thirds of New Zealand was once covered in native forest.

(Page 32)

As a nation we have succeeded in retaining our most important protection forests but their effectiveness for protection has become lessened and complicated by the effect of introduced animals in them. The photograph shows the lower interior of a rata-tawa forest (*Metrosideros robusta—Beilschmiedia tawa*) undamaged by introduced browsing animals. The floor is thickly covered with ferns and seedlings of trees and shrubs, the lower storey is well occupied. This was a normal interior of this type of forest but it is difficult to find now because of the browsing and trampling effects of animals.

(*Page 34*)

This mountain beech forest on the Craigieburn Range of the dry, eastern Southern Alps is in the process of disintegration. The alpine vegetation has been destroyed by fire, sheep, red deer, and chamois. Erosion is taking place inside the forest, now the harbour of browsing animals.

(Page 34)

increasing populations gradually carved out farm lands, but part of most farm holdings in Europe and North America in particular were left in trees to provide the essential needs—mainly firewood at first—of the farm. These farm forests, because of their large aggregate area, gradually gained importance in supplying larger and larger proportions of the raw material to the developing forest products industries. Improved management has led to greatly increased yields. The control of these yields and the fate of the forests themselves have come within national forest policies.

European immigrants (mainly British) to New Zealand found forests a great deal more complex than the relatively simple coniferous or deciduous hardwood forests in their homelands. For example, throughout Scandinavia there are only half a dozen main forest trees and three of them supply most of the forest produce; these forests have little undergrowth or ground vegetation. In New Zealand a most favourable growing climate and the complete absence of browsing mammals permitted the evolution of mixed evergreen hardwood-coniferous forests which made the utmost use of the sites offered to them. Closed-canopy forest with multiple storeys, densely carpeted floors and a tangle of climbing plants and lianes were normal.

Although the new settlers had reason to be glad of the ready supply of firewood, fencing posts, and some of their building wood, the forests impeded farming (apart from the browsing of free-ranging domestic stock) and held few prospects of management for permanent wood supplies as did forests in the Northern Hemisphere. Moreover, cleared and newly burned forest proved to be so fertile for grass and clover swards that settlers endeavoured to clear all forest as quickly as possible. Even in the kauri forests, in which there was promising potential for management and which grew on infertile soils difficult to farm, millers were allowed to dominate the scene and fires destroyed whatever forest remained after logging.

The developing young country thus turned a great amount of energy to the clearing of forest. The goal was profitable agriculture, and settlers were unfettered by rules and laws of land use based on experience as in Europe. With the help of axe and fire, clearance of forest proceeded rapidly and thoroughly. Many became alarmed at the rate, and their concern had the effect of coordinating into a strong group those concerned with the conservation of nature. Hence there came early the setting aside under various Acts of reserves called climate reserves, scenic reserves, domains, national parks and so on.

Bell arrested the clearance at a critical stage; continuation at an unabated pace would have seen the liquidation of native-timber resources before sufficient exotic forests could have been grown. Substantial inroads had also been made into true protection forest, but Bell halted this too by establishing most of the protection forest as permanent State forest, a restriction which the would-be clearer would find difficult to circumvent. Repeated attempts are still being made to disestablish these reservations.

Protection Forests Remaining

The essential function of the remaining protection forest is now clearly recognised and provision is made for the safeguarding and management of it under the National Parks Act (1952), the Forests Act (1949), and the Soil Conservation and Rivers Control Act (1941). Europeans, even those who arrived early, must have quickly recognised the importance of this forest. The large proportion of high country—much of it exceedingly steep—and the number of rivers with steep gradients and subject to frequent flooding, pointed to the function of the forest cover as the stabiliser of the mountain soils and the regulator of water. Under the State Forests Act, 1885, therefore, provision was made for State forests to be divided into three classes, one of which was called "mountain reserves" comprising forest reserved for protection purposes. In such forest timber was to be felled only by special selection.

No further attempts to define or to reserve protection forest by legislation were made until Bell's day. Although the safe limits of clearance were overstepped in many places, Crown ownership of the forest prevented wholesale clearance. As indigenous timber resources have receded, however, logging has been extended to higher and higher altitudes to find millable trees. Though only scattered, marketable trees are extracted and a matrix of hardwoods remains, even limited removal of trees from steep country has alarmed people living in districts in which unruly rivers flow from the mountains.

Apprehension was greatest in the 1950s, when large-scale logging began in the Maori-owned forests of the Urewera highlands. In this area rise rivers, subject to frequent flooding, which flow through the valuable Bay of Plenty farm lands, and the Urewera forests are of paramount importance in providing some measure of river control. Because of the weight of public opinion against logging, the Government decided to use the legislation of the Soil Conservation and Rivers Control Act, a section of which is as follows:

"32. Safeguards publicly notified—(1) The occupier of any land in any catchment district or catchment territory shall carry out every operation affecting the land in such manner and by such method as will conform to prudent land use practice, being practice which has proper regard to timing and circumstances and is likely to prevent so far as it is economically practicable, or (if prevention is not economically practicable) likely to mitigate soil erosion, and likely to promote soil conservation, the avoidance of deposits in watercourses, and the control of floods.

"(2) No person shall, without the consent of the Catchment Board or the Catchment Commission or the Council, as the case may require, do on or in respect of any such land any act or matter or thing which that Board or Commission or the Council has, by notice publicly notified within the immediately preceding two years, declared to be likely to facilitate soil erosion or floods or cause deposits in watercourses."

Under this legislation a form of consent was issued for the felling of standing timbers from certain areas which provided for:

"(I) ensuring the soil stability and the water regulating qualities of the said block;

(II) safeguarding the said block from fire;

(III) safeguarding the said block from any other damage of any nature whatsoever."

The construction of roads and method of felling trees was also to be supervised.

The exploitation of forest which serves the function both of protection and of production is an important issue in New Zealand because most of the remaining native forest is on steep country. In Europe this problem has been solved by controlling the activities of private people owning forest of this type and by carefully controlled management of State forests. This is the solution arrived at for the logging of the Urewera forest and the same formula will, no doubt, be applied in time to other parts of New Zealand. Foresters maintain strongly that forests should be managed for multiple-use purposes; that is, a forest can provide protection to soil and water, yield produce, provide recreation and serve other functions. The dominant use of any forest can be laid down by plan and other uses made subservient to the main one. In this regard the arguments that revolved round the fate of the Urewera forests holds several lessons for New Zealand.

Protection Forest and Introduced Animals

As a nation we have succeeded in retaining our most important protection forests but their effectiveness for protection has become lessened and complicated by the effect of introduced animals. Altogether some sixteen species of browsing and grazing mammals—excluding escaped domestic stock—have become acclimatised in forest and the associated protective vegetation of the mountains.

Captain Cook noticed the absence of native mammals during his first visit and considered that wild animals which could be hunted would be a useful source of food. He therefore liberated pigs and goats in the Marlborough Sounds during his second visit. (Also two sheep, but these soon died.) Settlers brought in rabbits from about 1840 onward and from then until 1910 many animals, some of which became successfully acclimatised, were introduced.

At first, additional food supplies were the aim of this activity, but this soon gave way to a desire to introduce animals for sport, for fur, as tourist attractions, or as predators to control animals already established. Acclimatisation societies, the Government Tourist Department, and private citizens all undertook these introductions; they were under no control until almost the end of the nineteenth century, when the consent of Government was required for the introduction of any animal or bird.

The most striking feature of all this activity was the remarkable speed with which some animals acclimatised and spread. Often spread was assisted by local "planting" of animals already established elsewhere in the country, an activity that still goes on despite its illegality. The red deer is an example of an introduced animal that quickly thrives and it has caused the most extensive modification of forests. A native of the wooded hills and mountains of Europe, it was first liberated in the Maitai Valley, Nelson, in 1851. About 30 further introductions from a number of sources continued until 1910. By the 1920s some herds had linked up so that deer occupied continuous, extensive stretches of country. Today they are present in mountains and forest (native and exotic) throughout at least half the country.

From about the beginning of this century there was growing uneasiness at the increase and spread of some animals. This was expressed in changes by legislation, by bounties paid by acclimatisation societies for shooting deer, and by many public expressions of alarm. In 1927 all protection was lifted from red deer in State forests, a recognition of the damage they were doing. Complaints about red deer competing with domestic stock for grazing in tussock grasslands compelled the Government to hold a meeting in 1930 to discuss problems caused by introduced animals.

After the 1930 meeting protection was lifted everywhere from introduced animals except opossums. (The taking of these was restricted until 1947, but by 1951 there was such concern about opossum damage that a bounty was introduced to encourage eradication.)

In 1931 regular Government control of deer was begun through a Deer Control Section set up in the Department of Internal Affairs. Legislative authority was provided in the Wildlife Act. Herds were still spreading (as, indeed, they still are) but little was yet known about the full impact of animals on forest, especially the combined effects of different species, though the severe depredations caused by large numbers of animals was obvious in many places. At worst, introduced animals could destroy forest and other mountain vegetation on steep slopes; everywhere they brought about minor modifications at least.

Control began mainly on grazing country leased from the Crown. In 1956 operations were transferred to the Forest Service and legislation (Noxious Animals Act 1956) was passed to provide the legal basis "to make provision for the control and eradication of harmful species of wild animals". Research work both on animals and the vegetation of the country they occupy was greatly increased to provide a sounder basis for attacking the intricate problems of control.

Ever since Government control operations (or culling as it was at first known) began there have been objections by sporting bodies and individual parties. Lack of effectiveness or of an adequate technical basis on which to carry out the work are the reasons usually given, but unquestionably these objections are based on concern about interference with hunting. Sportsmen have also sought to have certain animals—slow spreading deer species—exempted from the Noxious Animals Act, and certain areas set aside for game management. On the other hand, many bodies advocate the eradication of all wild animals, but this is practicable only in limited areas.

The clamour from sportsmen has sometimes induced governments to call widely representative meetings similar to that held in 1930. On the last occasion the issue was the management of wapiti deer in Fiordland National Park. This herd, occupying part of the park, is highly prized for trophy hunting. The Government finally instructed the Lands and Agriculture Committee of the House of Representatives to examine all aspects of noxious animals control. After exhaustive inquiries taking two years the committee was completely convinced that animals caused widespread damage and it recommended that there be no weakening of the Noxious Animals Act.[1]

Another of its many recommendations was that the farming of deer in strict captivity for the production of venison be permitted. The shooting of red deer for the export of venison has increased rapidly in the past few years. It can reduce deer to low numbers in country from which carcasses can be taken out economically. Helicopters and jet boats are being used extensively to get hunters into remote areas and for rapid recovery of game meat.

Remaining Indigenous Forest

In 1955 a national forest survey of indigenous forest was completed by the Forest Service.[2] Apart from giving detailed information about our forests it revealed that the total extent of indigenous forest was about 14.5 million acres or 22 per cent of the land surface. Of this about 2 million acres could be considered as exploitable forest by present standards of utilisation, and the remaining 12.5 million acres are primarily protection forest. Broad details of ownership of this forest are set out in the following table.

---

[1] Report of The Lands & Agriculture Committee on Noxious Animals Control. Govt Printer 1965.

[2] The National Forest Survey of New Zealand, Vol. 1, 1955. The Indigenous Forest Resources of New Zealand. Masters, Holloway, and McKelvey. 1959.

NATIONAL SUMMARY 1955

Extent of Forested Land   *Thousands of acres*

Class of Forest	SF	UCL	F & L	ML	RS	Total	Available
Merchantable	1,300	150	350	260	180	2,240*	2,060
Logged	370	40	220	100	---	730	---
Other	5,300	1,400	1,600	800	2,400	11,500	---
Total	6,970	1,590	2,170	1,160	2,580	14,470	2,060

*Softwood, 37%; mixed, 43%; hardwood, 20%.

(SF, State forest. UCL, Unoccupied Crown Land. F & L, Freehold and Leasehold. ML, Maori Land. RS, Reserves)

Notes:
(1) A possible 200,000 acres of logged forest is included with "Other Forest".
(2) There is an additional 1,750,000 acres (approx.) of land with State forest as legally defined: non-timbered highlands, swamps, riverbeds, etc.; a further 821,000 acres is State exotic forests.

Development of Exotic Forests

For several decades after the beginning of organised settlement, exotic forest trees were introduced in trial plantings. The early part of this period coincided with considerable European-organised plant collecting in western North America, Africa, and Asia. The first great journey of plant hunters was that of Francis Masson, the first official collector from Kew Botanic Gardens in 1772. There developed a veritable fever for the discovery and acquisition of foreign plants. Kew and the English Horticultural Society sent out more than twenty collectors before 1843, and others went on behalf of private individuals and horticultural businesses. Douglas, Don, Lobb, and Fraser are well known and several plants have been named after them. Nursery catalogues of about that time list many forest trees introduced to Europe from various parts of the world. One, issued by Conrad Loddiges & Sons, Nurserymen, Hackney, in 1836 contains *Pinus insignis* (now *P. radiata*). It is quite possible that this native of California and other trees found their way to New Zealand via England. Catalogues issued by the first nurserymen to set up business in New Zealand listed many recent tree introductions from the New World to the Old. Soon afterwards the discovery of gold in 1862 brought a rush of miners, some of whom might easily have brought tree seeds with them directly from North America. The very early connections and trade between Australia and New Zealand also resulted in the introduction of many eucalypts, wattles, and other Australian trees.

By these and by other means numerous exotic forest trees were introduced at an early stage of settlement and planted for shelter or ornament or in plantations. Some seem to have become acclimatised quite soon, among the most notable being radiata pine. When the first tree-planting encouragement Acts were passed soon after intensive settlement began this was the tree planted extensively, particularly over the Canterbury Plains. It was the one used for planting when the Canterbury Provincial Government made grants of land to counties, and it was used also for shelter planting. Later, the plantations established by the counties under this scheme were put under the control of the Selwyn Plantations Board, which undertook what seems to have been the first organised forestry in this country.

When the fledgling State Forest Service formulated plans for great expansion of planting there were still large tracts of easy country which, for one reason or another, were difficult or uneconomic to develop for agriculture. Afforestation turned to these. The most suitable, from many points of view, were the extensive tussock or scrub-covered rhyolitic pumice areas of Rotorua-Taupo and adjoining areas. "Bush sickness"—stock ill thrift caused by the (then unidentified) lack of cobalt in the soil and hence in pasture—had prevented anything but the most desultory development of livestock farming. Until a remedy for the stock sickness was discovered in 1937 afforestation had virtually unobstructed progress in this region. It was looked on as the heaven-sent opportunity for this land with its mile upon mile of golden tussock, its easy slopes, and the great depth of free pumice ideal for tree growth.

The State Forest Service began to plant up the Kaingaroa Plains with great vigour, although originally the land had been bought for agricultural development and settlement. At the same time many afforestation companies acquired tracts of pumice land and planted it, eventually matching the area established by the State. The planting boom lasted from 1925 to 1936, by which time the area of Kaingaroa Forest planted had risen to 255,000 acres, and almost 170,000 acres had been planted by the largest afforestation company, New Zealand Perpetual Forests Ltd, later to be named New Zealand Forest Products Ltd. Thirty to 35 years later the large forest products

Part of the Naseby State Forest, Central Otago. The first planting by the Lands Department towards the close of last century was done mainly in such treeless, tussock grassland areas. The principal species chosen for the cold dry climate at an elevation of 2,000 ft, were ponderosa and Corsican pines.

(*Page 14*)

Golden Downs State Forest Nelson. The large-scale planting era of 1925–35 saw the start of modest planting on the Moutere gravels, Nelson, which were only marginally suitable for farming. The forests there are being expanded to provide eventually raw material for a pulp and paper industry. Plantations can be seen on the Moutere gravels and remnants of beech forests on the mountains behind.

(Page 39)

units started to draw their raw material from these forests, planted mostly in radiata pine.

The Forest Service and private companies planted in other areas throughout the country, most extensively (but far below the pumice-lands scale) on the Moutere gravels, Nelson, a semi-consolidated boulder formation of steep and irregular terrain stretching from the north bank of the Buller River to Tasman Bay and covering about 350,000 acres. The coastal end of this formation was being developed in the 1920s for the growing of fruit trees, but the rest was mostly scrub and gorse-covered land, the scrub apparently having been induced by the burning of natural beech forest in Maori times. This was favourable country for the growing of exotic trees, so the Forest Service started Golden Downs Forest, and several companies small forests.

Other State forest plantings were:

Riverhead, just north of the upper reaches of Auckland harbour, on difficult and desolate gumlands[1] covered by stunted native scrub and introduced weeds—an aftermath of kauri forest (which brings about podzolisation of soils), fires, and kauri gum digging.

Maramarua, waste scrubland, south of Auckland.

Karioi, tussock grassland on the lower southern slopes of Mount Ruapehu and, at the time planting was begun, well above the altitude of intensively developed agriculture.

Eyrewell and Balmoral, on the shallow shingle soils on the north banks of the Waimakariri and Hurunui Rivers in North Canterbury, areas at that time considered too poor for agriculture.

West Coast, on the cutover terrace forest.

The discovery of a cure for bush sickness changed, virtually overnight, the prospects for agriculture in the pumice lands. It led to an era of large scale breaking-in of this land for agriculture by the Department of Lands and Survey, followed by subdivision and settlement by farmers, especially for ex-servicemen of the Second World War. Prospects of further large-scale afforestation on these lands disappeared. Indeed, such was the area already occupied by exotic forest that before industry

---

[1] Stiff clay soils in north Auckland that once carried kauri forest which left kauri gum in them. This gum was dug for extensively, leaving the country pockmarked by waterlogged hollows. Most gumlands have now been converted to agriculture or exotic forests.

became established to use it, various factions turned their thoughts to replacing the trees with pasture, and indeed some of the first areas felled were so converted. The massive forest products industry has changed the scene to such an extent that the established forests are now insufficient for its future needs even when intensive management increases their production. Industry looks with covetous eyes on agricultural land adjacent to its forest boundaries, just as do farmers at the good land under forest. Both forms of land use have their places in pumice country that is still to be developed, particularly so because erosion has developed in some areas indiscriminately sown to pasture.

The era of extensive afforestation, therefore, saw the planting of large blocks in the pumice lands, the start of modest planting on the Moutere gravels, and establishment of other forests scattered throughout the country. All lay in areas outside the boundaries of farming at the time they were planted. Before long, however, intensive livestock farming reached their boundaries.

Towards the end of the Second World War, the State contemplated the acquisition of further areas for afforestation to provide interim employment for returned soldiers until job opportunities opened up with the resumption of peacetime activities. Though this need never arose, new areas were acquired, some scarcely marginal to agriculture but well located for the supply of timber to markets. They provided a number of new foci for afforestation. Glenbervie lay almost at the back door of the thriving town of Whangarei. Rotoehu, in the Bay of Plenty, was well placed since, in time, it could supplement supplies from the large forests further inland. Two forest areas were acquired in the hinterland of Hawke's Bay, Gwavas and Esk, the latter at the head of the Esk Valley, where severe flooding and silting occured in 1938. Another area was at Patunamu, not far from Wairoa. An earnest start was made to develop afforestation in the Wairarapa by the acquisition of Ngaumu, east of Masterton. The long-term aim was to add to this forest, eventually to supply the Wellington market, which has no large exotic forests nearby. In the Nelson-Marlborough region, an area in the Rai Valley, between Nelson and Blenheim, was acquired, thus creating a resource additional to that of the Moutere Hills. Ashley, on the foothills of the Canterbury Plains, was taken over partly as a gorse-control project; it was well situated to supply Christchurch industries. A small area at Omihi in North Canterbury was taken over primarily for the control of a severe infestation of *nassella*[1] tussock, an aggressive introduced weed grass. In the south, areas were acquired near Dunedin on the hills surrounding the Taieri Plains, and in west Southland on the edge of beech forests which were being exploited.

This series of forests together with those already established, provided a reasonable distribution of exotic forests throughout the country for regional timber requirements, and, as was soon to be proved, the central North Island ones provided for the establishment of large industries.

By the mid 1950s the older exotic forests, principally of the radiata pine planted in the boom years, were being exploited on such a scale, and the produce used to such an extent by industry, that the success of exotic afforestation was well on the way to being assured. There had been mistakes: species had been planted which later proved unsatisfactory, or they had been planted on unsuitable sites. Some wrong decisions were to be expected because much of the knowledge of afforestation had come by trial and error over a long period. The most remarkable feature was that, despite failures, so many species of forest trees grew so well over such a wide range of soil types and climatic conditions. More serious mistakes arose from ignoring forestry principles: by planting large tracts to one species and within a short time, thereby creating considerable harvesting difficulties later. Nevertheless, a very large amount of utilisable growing stock was established.

Planning for the Future

The National Forest Survey that had examined the indigenous forests was extended between 1959 and 1962 to exotic forests and recorded, approximately, their size and location throughout the country. Investigations were also begun on likely forest-products requirements for the domestic market to the year 2000 and even beyond.

These calculations gave a reasonably firm basis for planning afforestation to meet future internal requirements for forest products, but it has become apparent in the past few years that New Zealand has such advantages in growing exotic forests that she can look to forest products as an important export. Markets,

---

[1] *Nassella trichotoma*, a grass introduced from South America as an ornamental plant. It is unpalatable and had taken charge of some areas of natural tussock grassland.

of which more will be said later, appear to be sufficiently promising for exports eventually to rank with other major exports such as butter, meat, and wool.

The Government therefore decided to adopt a policy of achieving a total area of afforestation throughout the country sufficient to yield domestic needs and a surplus of wood to supply export industries. This surplus was to be of the order of 150 million cu-ft[1] of wood per annum. The implications of this figure will be dealt with later. It is sufficient to say here that it was proposed to raise the exotic forest estate, which was just over one million acres at the time of the survey, to two million acres by the end of the century and to three million acres by the year 2025. Planting by the State was to be adjusted so that it would have planted about 600,000 acres of this target by the year 2000. It was hoped that the remaining 400,000 acres would be planted by companies, local bodies and individuals. Recent developments—such as the limited free trade pact with Australia and devaluation of New Zealand currency—indicated that these targets needed to be raised and they have been considerably. Programmes of this magnitude demand a more critical examination than has been made hitherto of the disposition of land between farming and forestry.

New Zealand has not yet quite reached the end of the era of forest and scrub clearing to make way for agriculture. And as agriculture has improved its techniques and gradually extended its boundaries many of the areas chosen in the past for afforestation could now make good agricultural land. Farming is a vigorous industry and contains some of the strongest organisations in the country, so that there is much organised opposition to the use of land for afforestation. Nevertheless, for various reasons, extensive areas would be better under trees. One is the steepness of much land that has been cleared. It was on this land that rivers, streams, and slopes developed their characteristics when under a forest cover. Removal of this has brought calamitous readjustments, especially where highly erodable rock formations have been bared. The constant burning of regrowth on some country has also encouraged widespread invasions of weeds. Both inside and outside the boundaries of agriculture there is much land suitable for exotic afforestation. In fact, a

---

[1] Cubic foot is a unit of true measure equivalent to a cube of 1 ft. side. In New Zealand, standing forest trees and logs are measured in cu-ft. As an example, the total amount of wood removed from New Zealand forests in 1968 was 222 million cu.ft.

broad analysis discloses that exotic afforestation, even on a much larger scale than that proposed, should not be short of land for a long time.

When the Government adopted a policy of additional afforestation, it also agreed that the planting programme would be organised to include: small afforestation areas whose produce would supply mainly local requirements; larger areas required to produce wood for forest products industries concerned primarily with export of commodities.

The industries based on the central North Island, though large by New Zealand standards, are already looking beyond horizons set by the present forest areas; they are therefore seeking land near their mills to plant for themselves. In addition, Government policy supports continued planting for these major industries. In the central North Island pumice region little land is now available in large blocks for planting, but there is ample room for trees in a readjusted land-use pattern that will allow both good farming and forestry. The main reason for seeking land close to industrial plants is that the cost of transport is one of the prime costs of getting raw material into plants—usually it is higher than the cost of growing wood. Forests close to plants are, therefore, very important as sources of cheap raw material. Inevitably, as they develop, forest-products companies will try to acquire for planting as much land as possible near their plants. There are already encroachments on to land that is being inefficiently farmed. Later, forest industries could compete for possession of more highly developed farm land on the basis of an economic comparison between farming and forestry.

Since planting on the Moutere Hills in the Nelson region was far greater than was needed for local wood supply and there was ample land marginal to agriculture, the Government decided to choose the Nelson area as the next to be planted up to support large-scale industry. Experience in the centre of the North Island suggested an adequate area of exotic forest would need to have at least 100,000 acres of radiata pine forest reasonably close to the proposed industrial complex. The total area of exotic forest would have to be much larger than the 100,000 acres, because not all sites would grow radiata pine. Moreover, for forest hygiene, large areas of one species should be divided by plantings of other species.

From the very brief New Zealand experience it would appear that large pulp and paper units are most conveniently sited on the coast near deep-water ports. Using the formula, then, of a forest or group of forests containing 100,000 acres of radiata pine and a utilisation unit near to a harbour, it would not be difficult to choose several suitable localities where forests would fit in to land use without encroaching much on established farming. Indeed, in certain localities forestry would bring considerable improvement in land use. The Otago east coast region is one, and it has already been nominated as a centre where planting will be built up to a large area.

Private planting looks mainly to radiata pine, while planting by the Forest Service is mainly of that species plus Douglas fir and Corsican pine. Radiata pine has become the staple species for industry, but the wood of the other two will readily be fitted into a pattern of use. The three species can be grown over an extensive range of sites where afforestation is likely to be carried out.

Throughout much of the tussock grasslands at lower elevations in the South Island conditions are not suitable for radiata pine or Douglas fir, but Corsican, ponderosa and lodgepole pine can be grown well. The last two produce wood prized by industry in their North American native habitats, and industry here would no doubt use them in time. The average per-acre return from pastoral use of tussock grassland country is very low: an average crop of trees would earn many times the amount, assuming of course, that industry was geared to take the produce. As a great deal of this tussock grassland has become degenerate —some even beyond the point of rehabilitation for grazing— planting of this country would serve a double purpose of production plus rehabilitation. Provided, therefore, the pulp and paper industry becomes firmly established and looks for increasing supplies of raw material, by far the greatest potential for production lies in afforestation within the tussock grasslands of the South Island.

Rehabilitation of Degraded Land

To complete the description of the pattern that exotic forests are likely to develop, more must be said about the rehabilitation of degraded lands. That they are present, extensive and in parts are severely degraded was one of the two main reasons for setting up the Soil Conservation and Rivers Control Council in 1947. Degradation will continue as long as certain types of steep country are occupied by stock.

Apart from that in some areas of high country in the South

Island, degradation is worst in the excessively shattered and soft rock formation of the Poverty Bay—East Coast region. Some erosion there is already as extreme as that found anywhere in the world. Afforestation, with the object of production as well as protection, is being undertaken by the State on one of the worst areas, Mangatu Forest in the Waipaoa River catchment. It should succeed in stabilising the land and also in producing good timber, because of the high quality of the soils. But bold national planning will be needed to save other large areas of similar country. A proposal for a very large afforestation scheme for eroding country along the east coast north of Gisborne has been accepted by the Government.

In coastal districts—mainly on the west of the North Island—there are over 300,000 acres of drifting sand loosened by grazing partly consolidated sand dunes which originally carried native vegetation. The drifting sand threatens farm lands in its path. It can be stabilised, provided the work is done methodically by forming a foredune, planting a complete cover of marram grass and sowing yellow lupin seeds. Such fixed sand cannot safely be grazed again; in fact, constant watch must be kept for breaks in the foredune or in the marram and lupin cover. In this stable vegetation, however, radiata pine can be readily established. To the windward, trees remain stunted and wind shorn, but to the leeward of this belt commercially valuable stands can be developed.

Sand fixation and tree planting was begun in 1922 at a few points on the west coast of the North Island by the Public Works Department (now the Ministry of Works). It was expanded when the Forest Service took over in 1951. As a matter of policy, afforestation of sand dunes is included as part of the State's programme. Some of these forests, because they are near markets, are expected to be most profitable in spite of the additional costs of sand fixation before planting and the need to maintain a purely protective belt next to the shore.

A few afforestation schemes have been started—where the soils will grow satisfactory stands of trees—partly to control serious noxious weeds. On some country too steep for implements to work, gorse has got a hold and frequent fires have induced dense cover. The only means of permanently suppressing this dense gorse is by planting a crop of trees or by waiting for it to be gradually supplanted by native vegetation. Farmers frequently plant radiata pine, because of its fast growth, to eliminate a "dirty corner".

A form of degeneration in an economic sense only is created when scattered timber trees are extracted from indigenous forest, leaving the secondary and uneconomic species behind. Though the altered forest is frequently essential for protection, it serves no economic purpose. It may grow on good soil. If it does, there is every reason to use that soil by converting the indigenous to an exotic forest, provided this does not reduce its protective value. In the past widespread attempts were made by the Forest Service and some other organisations to grow exotic trees in the gaps left when commercial trees were extracted. What the forester terms shade-tolerant trees such as Douglas fir and Lawson's cypress were planted in the openings. It was found, however, that such introductions into native forest required constant and skilled releasing from competing vegetation and tending, and were subject to depredations from native insects, fungi and introduced animals.

Few such plantings were a success. The method used now is to clear the cutover forest completely and reforest to continuous stands of exotic trees. It can be expected that eventually several hundred thousand acres of native forest cut over in the past or yet to be cut over will be converted to exotic forest.

Future Pattern of Farming and Forests

The general pattern of remaining indigenous forest is unlikely to alter, except for comparatively minor changes round the margins of the main protection forest and clearance of some commercial forest still occupying easy topography. This is a continuation of the process of forest clearance to make way for farming that has been going on for 150 years. But it can be expected that clearance in future will take into consideration protection values and the fertility and nature of soils. Of the present 14.5 million acres of indigenous forest all except about two million are wholly or primarily protection forests. It is within the two million acres that there will be changes, either by clearance or introduction of management for production or by conversion to exotic forest. The protection forest is safeguarded as State forest managed under working plans under the Forests Act or as national parks, scenic reserves, or unoccupied Crown land.

Exotic forests and woodlots occur now as part of the land-use pattern of farming and forests occupying lowland and montane areas. One group of large forests in the Rotorua-Taupo-

Apart from some areas of high country in the South Island, the worst degradation is occurring in the excessively shattered and soft rock formation of the Poverty Bay-East Coast region. Some erosion there is already as extreme as any found in the world. An incipient slump is illustrated; its size can be gauged from the cattle on the right. Pastures are very productive until they disappear when the ground slumps.

(*Page 45*)

An advanced stage of slumping. The aggradation in the stream bed is probably over 100 ft deep. Mangatu State Forest, Waipaoa

Degraded lands used for afforestation. Forest has been burned on these hills, but second growth has beaten the sown pastures except in small pockets. Fires occur frequently in this type of second growth and a fire pattern can be traced in this photograph.
Te Wera State Forest, Taranaki.

(Page 45)

Sand dune reclamation planting of radiata pine at Waitarere on the west coast, Wellington district. The planting has been inadequately maintained and sand drifting from the coast has destroyed some. It is essential that the foredune be maintained and the sand kept fixed. The trees farthest from the coast yield good timber, and the sandy-soil farmland behind the plantation would be more profitable under trees.

(Page 45)

Tokoroa area, occupying over half a million acres, will expand as a result of deliberate Government policy, and also as a result of competition for land between farming and the forest products utilisation companies. This will be followed by developments in another large area in the Nelson region, already being planted. To this might well be added later the Marlborough Sounds area, and parts of the Waiau River catchment, in Marlborough, the vegetation of which is largely degenerate as a result of unsuccessful farming and constant firing. Because of the steep topography and lack of reliable natural water resources for stock the Sounds area land is difficult to rehabilitate for farming, but it is good tree-growing country.

Depending on the success of industry and its quest for further afforestation, large aggregate areas of forest could be planted elsewhere. Individually forests might not be as big, but the total area in any one region could be. Government policy mainly will dictate where these areas will be and at what rate they will be developed. The number of smaller forests—State, company, and municipal—which now occur from North Auckland to Southland will be added to in area and number.

Well-tended and productive woodlots on farms are still rare. Those that are established close to markets, however, show how profitable they can be in using areas of farm land of low agricultural productivity to grow trees. In this direction New Zealand has great potential for the growing and careful tending of high quality forest produce ranging from posts to veneer logs. In the hopes of getting farmers to use more of this potential the Government introduced a loan scheme in 1961. Loans help to finance the growing of trees to the utilisation stage. The main momentum to farm forestry, however, is likely to come from an improvement in stumpage (price for standing timber), which means that farmers must learn to market timber efficiently as well as to grow it efficiently. The forestry encouragement loan scheme has been extended to include local bodies, and taxation incentives were provided to encourage planting by companies.

An increasing part will be played by trees and forests in rehabilitating degraded lands. In the more distant future possibly the largest area of exotic forests will be planted in such country. Some of the planting must be purely for protection but most of the land, even though degraded, has soils that should produce raw material economically. Afforestation of sand dunes, planting for the stabilisation of eroding soils—immediately those of the soft tertiary rocks of the North Island, and in future the more severely eroded soils on the eastern side of the South Island mountains—and planting for the suppression of noxious weeds all offer considerable scope for the production of large quantities of wood.

Though in New Zealand the State will have to do nearly all of this planting, there is scope for small areas to be planted on degraded soils in otherwise good farm land. Willows and poplars (including some species with much better timber than the common Lombardy poplar) are used widely to stabilise such country, especially in projects subsidised by the Soil Conservation and Rivers Control Council. The planting of forest trees on a small scale is also subsidised, and on a large scale—up to 100 acres per annum—where planting will achieve both on-site and off-site benefits.

5

THE DEVELOPMENT OF FOREST MANAGEMENT

New Zealand is frequently criticised by overseas visitors, and New Zealand foresters by their own countrymen, for not making greater attempts to practise more of their forestry with indigenous timber trees. Because of this lack of indigenous forest management and the fear that the sawmilling industry would eventually destroy all indigenous production forest and even eat far into protection forest, there have always been strong movements to lock up native forests in reserves.

Scope for Management of Indigenous Forest

In many areas within its range the kauri, for example, could have been perpetuated in managed forests. Soils on which kauri forests grew for a long time proved difficult to convert to agriculture (and still present problems), and even when converted many of them were marginally economic. Some areas of forest might well have remained for the permanent production of this superb timber. The tree regenerates readily where there are seed trees and it grows at a reasonable rate, though not nearly as fast as the main exotic trees now used for afforestation. Although limited areas of kauri regeneration have been saved, most of the forests were destroyed in the process of logging and by the fires that followed, and no Government measures were ever introduced to halt this destruction. In fact the reverse could be said. As far back as 1877 the first Conservator of Forests, Campbell Walker, stated:

> "Nor must it be supposed that, although the supply of timber for the present is ample, it is by any means inexhaustible. No forest is inexhaustible unless systematically worked on principles which ensure the capital not being trenched upon and the income alone utilised; and, in the case of the valuable kauri forests of the North, the date at which

this exhaustion or annihilation will have become an accomplished fact may almost be set down as within the present generation. With a large export, both intercolonial and foreign, great waste in what the French style 'exploitation' and conversion, and no attempt at reproduction, Nature's efforts at which are frustrated by fire, the end is not far to seek. Kirk puts it down at forty years, and I am not sure that he is not beyond the mark."[1]

An accusing finger could be pointed at past generations of New Zealanders for having permitted this destruction of the kauri forests. But to have introduced management into other forests containing the main softwood timber trees would have been impossible, except in the rimu terrace forests of Westland.

Kahikatea, which was milled and exported in large quantities, had a special use for butter boxes because the timber was odourless. It grew in deep, fertile silt swamps, which were some of the first soils to be cleared and drained and sown down to pastures for dairying. Perpetuation of kahikatea could not be contemplated, good as its timber is, because of its very slow growth and demand for the best soils.

Rimu, matai and totara all grow on soils that have, in the main, been sought out by agriculture and, over the years, have been converted to farm land as the trees were felled for timber or simply burned to clear ground. It has been argued that areas of logged forests planted in exotic species or part of them should have been planted to rimu, and that the remaining rimu forest should be so managed as to reproduce rimu. However, about 300 years, possibly longer, are needed to grow a rimu tree of acceptable milling size. As this species—and matai or totara—would have to be grown on good soils, there would be persistent attempts by the agricultural community to take the land from forestry if it had to contemplate trees occupying good land for three centuries before yielding any return. Nationally such unprofitable use of good soils would be unacceptable.

West Coast Rimu Terrace Forests

The rimu terrace forests of the west coast of the South Island offer almost the only prospect of management of rimu for perpetual supplies of timber. These water-washed, glacial-boulder terraces have an impervious soil almost like plasticine,

---

[1] 1877 Report of the Conservator of State Forests. Capt. I. Campbell Walker.

below which is a series of dense iron pans. The soils have never been converted to agriculture in spite of the easy topography, nor does this appear likely in the 150-300 in. rainfall climate. They do, however, carry a good rimu forest, which has slowly invaded the terraces since the last glaciation. Mature trees are smaller than mature rimu growing elsewhere, but the forest is uneven-aged and regeneration occurs under certain conditions. Since at least 50,000 acres of this type of forest remain, it is legitimate to place it under management even if the rotation has to be of the order of 300 years. As the soils cannot be used for other purposes, it could be considered a grave error for a forest authority not to attempt management.

Until about 15 years ago these forests were opened to industry on the same basis as other native forests in New Zealand. Logging virtually destroyed the forest cover including young growth, and what was left was soon killed by fire. Semi-bogs, formed on the impervious soil, then took the place of thriving forest. Desolate-looking bogs, all the way from Westport to south of Ross, are the legacy of this type of exploitation. To use the resources of the soil in such a manner is, indeed, a blot on the record of the nation.

As a matter of national policy, the resources of Westland have been reviewed by the West Coast Committee of Inquiry set up by Government with the object of using them more effectively. Forests obviously had to come in for close scrutiny, for the forest cover of the Westland and Buller Counties amounts to about 50 percent of the land area. Most is protection forest in the steep, high-rainfall western flanks of the Southern Alps, but the production forests contain by far the largest area of New Zealand's remaining indigenous resource and must therefore be regarded as a national resource.

Apart from timber, Westland's earnings come from a limited amount of farming practised on the recent alluvial soils of the large rapidly flowing rivers, and from coalmining. It is agreed by all that, given the right conditions, per-acre production on farms could increase markedly, but there is not much scope for expanding the area on which it can be practised profitably. Coalmining over the past decade has declined. It is apparent, therefore, that the forests should be managed to play as large a part as possible in maintaining or, if possible, expanding the economy. This policy was, in fact, adopted by the Government in keeping with the findings of the Committee of Inquiry.

In implementing this policy, the Government has to recognise the increasing national importance of the indigenous timber supplies of Westland, quite apart from the place forests and forest industries will play in the local economy. Thus, long-term cutting rights are being granted to millers who improve the efficiency of their sawing and do processing in Westland. (Until recently most of the timber has left the district in the rough-sawn state.)

Management in the terrace rimu forests had to be preceded by the investigation of some difficult technical and economic problems. First and foremost fire has had to be kept out of logging areas. This has largely been achieved, mainly because of the change from steam locomotives to diesel-powered haulers, and to timber extraction over roads instead of over tramlines. Improvement in access through roading has improved the prospects of management. Selective logging has been introduced in the uneven-aged forests, but there are still problems. They include the formation of a close network of extraction roads, the adoption of logging techniques quite different from those at present used in clear felling, and the extraction of part of the crop only without damage to the rest. The changes might seem to amount to a revolution for local industry, but the techniques are part of normal forest practice in many parts of the world. The choice lies between selective logging or destruction of the forest and the soil; between managed forest or perpetual bogs.

Though, because of the slow growth of rimu, management of forests will eventually produce not more than modest—though sustained—supplies of timber, Westland industry will expect to expand. One of the policy measures that had to be adopted therefore was to start afforestation with exotic species so that a growing stock could be built up for the time when industry would begin to look for more wood. Exotic afforestation was attempted on the West Coast during the planting boom of 1925 to 1935, but as difficult terrace soils were chosen it largely failed.

For several years attempts have been made to afforest the hill soils which cannot be developed for agriculture after logging because of the luxuriant second growth that comes away rapidly in the wet climate. But the soils have very much better texture than the terrace ones, and techniques have been worked out to clear them and burn the vegetation in preparation for planting. So far radiata pine and Douglas fir planted on them are thriving. Successful establishment of exotic forests holds out the only hope of maintaining forest industries on the West

Coast on any scale. If they fail, only minor industry can survive, and there would be great difficulty in holding the manageable terrace forests against an industry that was running out of resources. The key to forest policy problems of the West Coast lies in exotic forests.

Kauri Forests

At the present time there is a limited area of kauri regeneration that escaped destruction during the logging of mature forest. A few areas of mature kauri also remain. Those not under permanent reserve (scenic reserves and forest sanctuaries such as the Waipoua Forest Sanctuary and Trounson Park) are being managed for a sustained yield of kauri timber—a small yield of between one and two million board feet per annum, but one that is likely to increase as regenerated areas come to maturity. The high priced timber itself is used for special purposes, particularly boatbuilding.

The present small yield is a far cry from that of the days when the cut of kauri throughout the North reached a peak of 110 million bd.ft and when an estimated 40 million bd.ft were exported. The area of merchantable kauri forest remaining is probably between 8,000 and 9,000 acres and the area of regeneration on State forests available for management is about 40,000 acres.

Planting of kauri has been attempted over many years but by and large without success. Good planting stock can be raised in nurseries and seedlings survive well when planted out. But though mature kauri stands occur on poor soils, planted seedlings grow satisfactorily only on good soils that are generally in use for agriculture.

When there is a source of seed from nearby trees natural regeneration occurs readily in nurse crops of manuka which have sprung up after fires. It is in these regenerated stands that management techniques are being worked out. They are concentrated mainly on the Great Barrier Island, near Auckland, in the area south of Russell, and in various places on the Coromandel Peninsula. Some have already yielded commercial thinnings, but it is expected that rotations of up to 150 years will be needed to produce good millable logs.

Southern Beech Forests

The southern beeches account for the largest part of the remaining native protection and production forests. They grow on soils which, when cleared, convert to medium quality or poor agricultural soils. For this reason fairly extensive areas of what could be considered as production forests escaped clearing and were made State forest soon after the State Forest Service came into being. Limited areas of these are in the North Island and more extensive areas in the south of the South Island, but by far the largest areas lie in the north-west region of the South Island.

New Zealand has three species and two distinct varieties—in effect, five species—of southern beeches, and there is a complex hybridisation between four of them. Only in the south are there large forests of one (silver beech) species.

Beech forests have a structure differing fundamentally from that of complex broadleaf-podocarp forests with their several storeys and tangle of low vegetation. Beech forests have one uniform canopy with only sparse vegetation beneath. Regeneration is prolific after heavy seeding years, so that any gaps in the canopy are quickly filled with beech seedlings and saplings, provided there is no interference by noxious animals. In structure and habit of growth, southern beech forests are akin to the Northern Hemisphere beech forests, in which foresters have developed exceedingly effective systems of silviculture based on natural regeneration. A similar prolific regeneration offers opportunities of evolving equally successful systems in the southern beeches.

However, expenditure to improve silviculture and management cannot be justified until there are good markets for beech timbers. So far markets have developed to a limited extent and only sporadically. The woods are hardwoods in contrast to the main supply of softwoods—at first indigenous and now exotic—which have been the raw material of industry. Each beech species has timber with markedly different properties and all the timbers are more difficult to saw, season, and market than are the very easily handled softwoods.

Markets have developed best for timber cut from the silver beech forests of Southland, where industry has had only this one species to saw and market. A system of management by natural regeneration has been started but techniques are far from being perfected.

In other areas, particularly in the forests of the north-west of the South Island, the development of beech timber marketing has been poor, because both at the sawing and at the marketing stages species have tended to be mixed and the presence of

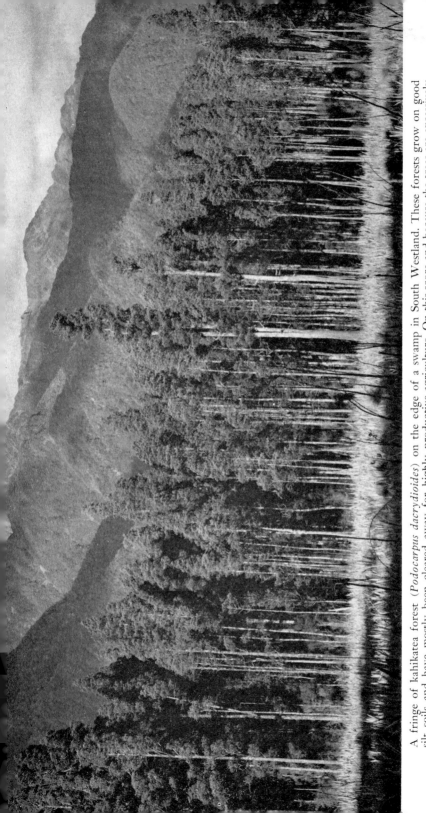

A fringe of kahikatea forest (*Podocarpus dacrydioides*) on the edge of a swamp in South Westland. These forests grow on good silt soils and have mostly been cleared away for highly productive agriculture. On this score and because the trees are excessively slow growing, these forests are not replaceable.

(*Page 50*)

The uneven-aged rimu terrace forests grow on washed glacial boulders overlain by a glacial till that has been laid by water. These are known as Okarito soils. Strong iron pans have been formed in the boulders below the till, which itself is like plasticine and impervious to water. This inhospitable soil supports good rimu forest, but has never been used successfully for agriculture or for exotic afforestation. Once the rimu forest is cleared a bog forms on the till.

(Page 50)

West Coast rimu terrace forests offer almost the only prospect of management of rimu for perpetual supplies of timber. The forest is uneven aged and regeneration occurs under certain conditions. Although the rotation will possibly be as long as 300 years the soils will, at present, grow no other crops.

(Page 50)

Kauri in Russell State Forest. Provided there is a source of seed from nearby trees, the natural regeneration of kauri can be readily achieved in nurse crops of manuka which have sprung up after fire. These stands are the ones in which management techniques are being worked out.

(Page 53)

hybrids has added to the confusion. A buyer therefore cannot be certain that he will get what he stipulates. Moreover, though a sawmiller might establish a market for the top grades of a reliably named timber, he is faced with the problem of marketing the low grades in competition with softwoods.

Because of the limited development of markets, little management has been introduced into beech forests. Nevertheless, the timbers are good and it would seem that the State might well make the effort of encouraging and guiding the development of their use. Perpetual supplies could be assured by the introduction of management of forests.

Recently, thoughts have turned to the possibilities of pulping beech woods. Trials of individual species have shown them to be suitable but any operation would have to deal with two or more species. If this is ever done, the forests from which the raw material is drawn would, in all probability, be clear felled and converted to exotic species. Southern beeches in general grow on soils that bear good exotic trees.

Of the major indigenous species, only kauri, rimu in Westland, and southern beeches have characteristics that enable them to be brought under permanent management by the forester. They are not easy forest trees to domesticate and, while a start has been made on this, success will come only after a long period of trial-and-error experimentation.

Other Native Trees

We can now examine briefly the prospects for permanent management of the other indigenous trees for which timber markets have been developed. In approximate order of present acceptability to timber merchants they are: tanekaha (*Phyllocladus trichomanoides*), tawa (*Beilschmiedia tawa*), silver pine (*Dacrydium spp.*), pahautea (*Libocedrus bidwillii*), rewarewa (*Knightia excelsa*)—not an impressive list, and one mainly of species secondary both in the forest and on the market. Those that are softwoods—tanekaha, silver pine, and pahautea—would undoubtedly be accepted in large quantities if the timbers were available.

Silver pine is by far the most durable timber we have and has been sought keenly for posts, poles, and railway sleepers. It grows in bogs, and on the West Coast trees that have fallen and lain in bogs, possibly for centuries, are salvaged. Unfortunately, it is exceedingly slow growing.

Tanekaha occupies sites which generally go into agricultural production, but it is also a component of the kauri regeneration areas, where it grows well even on poor soils. It seeds at an early age and prolifically, so that in time a supply of tanekaha logs should be obtained from these predominantly kauri areas.

Pahautea grows on ground outside the range of agriculture, but forests in which it occurs are limited and often isolated, and its growth, like that of silver pine, is excessively slow—30 rings per inch, or a twelve-inch-diameter log produced in 300 to 400 years.

Forests containing tawa are fairly extensive in the centre of the North Island, and a concerted effort has been made, with success, to saw and market the timber. The tree produces high grades excellent for flooring and furniture, but it is very susceptible to the common house borer and an essential part of the drive to get the timber established is its compulsory preservative treatment. Some studies have been made on the management of tawa forest and these have produced hopes for success. It grows on soils of agricultural value or, if too steep for this purpose, on soils that will grow good stands of exotic conifers much faster than it will grow tawa. It seems that in the long run tawa management will be confined to limited areas of steepish country. Possibly its management will extend gradually into forest now considered as purely protective. Much of this type of forest contains good stands of tawa which, with care, could be logged without damaging the remaining forest trees.

Many sawmillers in the North Island have cut rewarewa, which has a limited market, again only if treated against borer attack. The tree regenerates readily and even prolifically in second growth, but the areas where this happens are likely to be converted to exotics. So far there has been no conscious management of the tree, though this is likely to be undertaken on small areas.

The foregoing discussion of the possibilities of perpetuating indigenous timber species explains why so little management of native softwood forests has been attempted. It would be impossible in New Zealand to devote large areas of good land to growing rimu or allied trees that yield only a few tens of cubic feet per acre per year when exotic softwoods yielding ten times more could be grown.

In contrast, the native hardwood beech forests grow on relatively poor soils and extensive areas remain. Trees grow somewhat faster than native softwoods and there is prolific regeneration. Some permanent management has been introduced and it is expected that it will be possible to obtain yields of between 80-120 cu.ft per acre per year, which are about the equivalent to those from European beech, extensive areas of which are managed in continental Europe.

Dependence on Exotic Forest Trees

Exotic forests in New Zealand present a very different picture, and State forests and most company forests have been planted with a view to giving sustained yields. We are just learning what these yields can be, how they can be increased by good silvicultural practices, and how the wood can best be sold to meet the needs of industry.

Well before the end of the nineteenth century it was clear to many people that eventually the indigenous timbers would run out.[1] Afforestation with quick-growing introduced trees therefore loomed up as an important policy early in the history of the colony.

That New Zealand received its share of plant introductions which came to Europe from the New World may be seen from plant names included in the catalogues produced by the first nurserymen in this country; they contain a surprising number of North American conifers. They came as seed from the original introductions to Europe, and as potted plants ingeniously kept alive during long voyages in the cramped and crude quarters of small sailing ships.

By the time tree-planting encouragement Acts were passed in the 1870s radiata pine seems to have been well known (though under the name insignis pine) and favourably regarded for shelter because of its fast growth. Only 25 years earlier it was a tree grown in pots and advertised by nurserymen at 10s. ($1) per plant; today it is sold at about $12 per 1,000 plants suitable for setting out. This introduction, of unlikely promise, growing in natural stands only on a small part of the coast of California, quickly gave indications of its later development as the main forest tree of New Zealand. Burning of scrub and second growth undoubtedly destroyed some early plantings, but it had the

[1] The permanent supplies available if sustained-yield management were introduced into selected indigenous forests would probably be something less than one hundred million bd.ft—possibly not more than fifty million bd.ft out of a total probable demand at the end of this century of well over one thousand million bd.ft.

compensating effect of opening up the normally closed cones—releasing seed and inducing regeneration. In this way the tree quickly became acclimatised.

In the more rigorous climate of the treeless districts of Otago and Southland, where shelter was also wanted, radiata pine could not be grown, but European, Scots, Corsican and Austrian pines, Norway spruce, and North American yellow pine (*Pinus ponderosa*) were planted. They proved to be much hardier, though much slower growing. However, Scots pine, the main timber tree of Europe, never thrived as a forest tree here—nor has it when introduced elsewhere in the world—and European and Norway spruce were soon attacked by aphis and could never be grown widely.

The Australian eucalypts were always attractive to plant introduction enthusiasts because of fairly regular early traffic between the two countries, the great range of species from which to select—possibly some 600—and the fact that they provided the main timber supplies in their native country. Of the many introductions made, some flourished, some failed, others grew well for a time only to become diseased as their own native insect pests were inadvertently introduced and became established. The Mediterranean maritime pine soon proved to be at home on the heavy North Auckland soils and, like radiata pine, regenerated prolifically after fire and quickly became acclimatised. Early planting under good conditions showed up the potential of many trees, although the translation of these results to forest growth under generally more difficult conditions had to be done cautiously.

By the time the Afforestation Branch of the Lands Department had got under way in the 1890s, a surprising amount of knowledge had been gained about introduced forest trees. The main species for the afforestation projects were selected soundly, and seed supplies were carefully secured from good sources. Remarkably few errors were made in plantation techniques. The main species used were the European trees Corsican pine and larch and the North American trees Douglas fir and ponderosa pine. At first only small plantings of radiata were made without, as one observer cynically remarked, any faith that the tree would be utilised. The timber at that time was virtually unknown.

A number of trees were planted for special purposes, the policy being to produce timbers that had to be imported and could not be replaced by indigenous timbers. Thus hickory was planted in the hope that tool handles would be obtained from it; equally popular was the North American catalpa, which produced very durable fence posts; there was a catalpa planting craze in the United States at that time and New Zealand followed the lead. In *Tree Culture in New Zealand*, written by the Chief Forester in 1905, a chapter is devoted to the tree, whereas radiata pine is mentioned only incidentally and is listed among trees suitable for shelter, though not among trees "Generally Recommended for Extensive Planting for Timber Purposes". Included here were Douglas fir and Corsican pine, the second and third choices now, and over 20 eucalyptus species.

Some errors in choice were inevitable, probably the most obvious being in the Australian eucalypts. The wrong species mainly were selected and seed was often obtained from poor strains. One species, *Eucalyptus globulus*, or Tasmanian blue gum, did well and the timber was readily millable, but the tree was soon attacked by scale and leaf-eating insects.

It did not take long to prove that the policy of growing trees for special-purpose timbers contained many fallacies. It was as well that New Zealand found these so early, because the policy is still advocated among those unfamiliar with the production and marketing of timber. Most special-purpose timbers are hardwoods, and those tried in New Zealand did not thrive. Moreover, the marketing of low grades of hardwood timber, some of which is contained in all logs, must be done in competition with the low grades of softwood timbers, which have a much greater range of uses and have always been produced here in abundance.

Many other species were planted by the Afforestation Branch of the Lands and Survey Department in small areas for observation and experimental purposes. Some that proved successful were *Cupressus macrocarpa*, *Chamaecyparis lawsoniana* (Lawson's cypress), *Thuya plicata* (Western red cedar) among the quality softwoods, and *Fraxinus excelsior* (European ash), *Quercus robur* (European oak), and various popular species on especially fertile or suitable sites. Diseases have subsequently attacked some of these. In addition, a great many species were planted out in arboreta: in fact, the number of species, and the range of countries they were introduced from, were remarkable.

By the time the State Forest Service was formed, the first plantations established by the Lands Department were about 20 years old and had reached young pole stages. About 40,000 acres had been planted, principally to the main species noted earlier. There were many examples of young stands planted on many different sites to guide the planting programme introduced with the new service. Thus the first two decades of the new century saw the genesis of an afforestation programme that capitalised on the success of forest tree introduction and growing on a small scale throughout the previous 60 years.

Two other British Empire countries in the southern hemisphere, Australia and South Africa, passed through much the same phases of exotic afforestation at about the same time.

Planting by the State Forest Service

Part of the policy of the new service was to expand the afforestation programme to replace the dwindling indigenous timber, and for this the Afforestation Branch of the Lands Department was transferred to it. The traditional building material, wood, was to be produced if at all possible. The circumstances for doing so were particularly favourable. No doubt MacIntosh Ellis, the newly appointed Director of Forestry, who was a Canadian, also had in mind the growing of wood for a pulp and paper industry, some knowledge of which he would have from his own country. By that time also, radiata pine had shown its real value as a rapid producer of wood, and its timber was known to some extent from the milling of the early plantings in Canterbury.

With this promising background and the careful start that had been made with State afforestation during the previous 20 years, a greatly expanded planting programme was got under way; but, regrettably, caution as far as forestry principles were concerned was now thrown to the four winds. There were radical departures from nearly all basic principles, and only fortuitous circumstances and exceptionally favourable conditions for tree growth in general gave New Zealand the successes it achieved. There were also failures of no mean magnitude—failures that are still appearing 40 years or more after planting.

Planting by the service increased from just over 1,000 acres in 1921 to close to 60,000 acres in 1929, but dropped again to about 2,000 acres in 1938. During the period 1924 to 1937, a total of 375,000 acres were planted in State forests; much the largest area of this was of radiata pine in Kaingaroa State Forest. The worst mistakes were:

> Planting of radiata pine on many unsuitable, particularly harsh, sites;

Not tending trees during the establishment stages, resulting in many poorly stocked and malformed stands;

Planting large areas at one time, so that when forests had grown to the utilisation stage too much was ready for logging at the one time;

Planting large areas to poor strains of trees, particularly the slow-growing *scopulorum* variety of *Pinus ponderosa* and poor forms of Corsican and lodgepole pines.

In spite of these mistakes and lack of later tending, radiata pine did succeed in many areas. The total effect of the boom was the creation of a tremendous volume of wood which, together with an equal quantity established by company afforestation, provided the raw material some 30 years later for large-scale industries.

Company Afforestation

Afforestation companies originated mainly from land speculation, though some, no doubt stimulated by the example of the State Forest Service, were genuine afforestation enterprises. Some of the land companies, too, fortunately the largest of them, put their afforestation houses in order once they hit on the combination of radiata pine and suitable growing sites.

Land-utilisation companies operated in many and varied ways. They bought waste land, which was abundant at that time, particularly the pumice lands about the centre and east of the North Island. Some then sold bonds, each bond providing for an acre planted to some crop and managed until the time it yielded produce, when bondholders would share the profits. Areas were planted to tung oil trees (in North Auckland), fruit trees, New Zealand flax (*Phormium.tenax*) and forest trees. This was the company promoter's heyday, for it seemed possible to conjure up investors from a great many sources, both within and without New Zealand. Even Indian princes were the unlikely target of one imaginative promoter. Tens of thousands were enticed into afforestation investment by rosy claims centred round the phenomenal growth of radiata pine and the early returns from it.

As was to be expected, some of these optimistic ventures failed in time, the tung oil companies completely. Almost 60 of the companies planted trees, some on quite a small scale, but others on a relatively large scale, representing bond investment running into millions of dollars. Only one company had an experienced professional forester on its staff, a few obtained professional advice, and some employed men who had gained experience in the Afforestation Branch of the Lands Department. But such was the speed of investment and the control imposed by their company-promoter masters, that even these men departed from the safe rules they had built up painstakingly by experience and so committed some gross errors. It can therefore be imagined how companies without any experienced men were likely to fare.

Quite apart from the technical aspects of afforestation, company promotion came under the searchlight both of public and of Government scrutiny. Often dubious tactics were used to sell bonds to people who had a very sketchy idea of what they were investing in. The promoters had little or no knowledge of what they were trying to do technically. Eventually a commission of inquiry* into afforestation-company and other company promotions led to the strengthening of company law to protect investors.

In plantings by the State on pumice soils and in company plantings radiata pine was the main success. Eventually about half a million acres, almost entirely of radiata pine, planted by the State and companies were successfully established. The peak year was 1929, when an estimated 99,000 acres were planted, but by 1939 the forestry company promotion bubble had been pricked and total planting that year dropped away to only 18,000 acres.

The forestry companies which set their houses in order soon began to formulate utilisation schemes. Others had to wait until their trees were large enough and markets were established before trying, under most difficult conditions of oversupply of wood, to sell stumpage. One company carried out the intention avowed in its original prospectus and began the manufacture of pulp and paper board, and later another started production of pulp and paper.

Thus, in the first 15 years following the formation of the State Forest Service there occurred the most important developments in New Zealand forestry: The beginnings of the creation by the State and forestry companies of very large amounts of raw material which, although not well grown, industry has found to its liking and is now using in many ways. What

---

* "Report of Commission of Inquiry into Company Promotion Methods, etc" 1934. Government Printer, Wellington.

Mature kauri in Waipoua Forest Sanctuary. Trees in the centre are possibly a thousand or more years old. The branchless, columnar trunks, and massive crowns are typical. Regeneration is to be seen in the disturbed soil on the roadside.

(Page 53)

A cedar forest (*Libocedrus bidwillii*). Cedar is one of New Zealand's minor timber species that would be used in much larger quantities if it were available. The timber is somewhat akin to the imported western red cedar. Forests grow at the higher altitudes on wet sites. Growth is excessively slow and trees in the photograph are several hundred years old.

(*Page 56*)

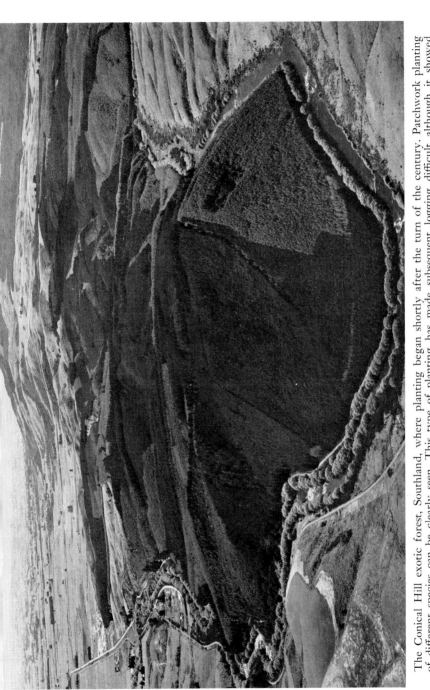

The Conical Hill exotic forest, Southland, where planting began shortly after the turn of the century. Patchwork planting of different species can be clearly seen. This type of planting has made subsequent logging difficult although it showed the vigour with which a number of exotic trees would grow. Both banks of the Pomahaka River flowing around the forest are lined with European crack willow. This is a common sight throughout New Zealand. (*Page 58*)

Radiata pine. An untended 35-year-old stand on a first-quality pumice soil site in Kaingaroa Forest. Some of the trees are over 120 ft high and have a diameter of more than 2 ft at breast height.

(Page 60)

happened in that period decided many aspects of forest policy that followed.

The era also taught many lessons that should not be forgotten. The planting of radiata pine in large, concentrated areas placed this forest tree, despite bad age-class distribution, firmly and solidly as the cornerstone of New Zealand exotic forestry for a very long time to come. When industry became established on the wood from these areas, the advantages of having available large quantities of relatively uniform raw material were quickly discovered. Cheapness of logging and of transport, conversion of large supplies of raw material and bulk marketing were of basic importance to industries and far outweighed any inherent disadvantages of wood from untended radiata pine stands. From being a minor forest tree, whose position could not be assessed before that period, radiata pine became the major tree. It continued to form a large proportion of the State plantings and became almost the only tree that companies would consider.

What tends to be forgotten now is that there were extensive failures of radiata pine where it was planted on sites subject to out-of-season frosts or on difficult soils. Degraded gumland clays in the north and granite soils near Nelson would not support thrifty growth of the tree. Some plantations were also inadvertently burned because sufficient finance had not been allowed for the protection and management of the large areas established. Finally the destruction by fire in 1946 of 30,000 acres of young radiata pine forest near Taupo jolted the nation into a realisation of the fact that in its exotic forests it had a valuable asset to protect. The comprehensive Forest and Rural Fires Act, 1947, passed after this fire, provided a good legal basis for future protection.

The establishment of large areas of even-aged radiata pine forest has had disadvantages and is likely to have many more before this position can be rectified in the second and succeeding rotations. There has already been one insect epidemic over a large area, fortunately well before utilisation had built up to the sustained yield of the plantations, so that the loss of wood was not of consequence. Other epidemics are certain to occur, and at times when the losses of wood may be serious. *Dothistroma pini*, an introduced needle-cast fungus, developed to epidemic proportions in 1966. The disease, which is described more fully on page 108, attacks radiata pine stands up to 15 years of age and ponderosa and Corsican pine stands of all ages. Over 80,000 acres of forest were sprayed in the summer of 1966–67 with a copper spray and the treatment has proved very effective for radiata pine.

A forest having a range of ages spanning only 10 years that must be harvested over 30 to 40 years presents the foresters with difficult management problems. It also poses problems for industry: the trees are growing all the time, so that machinery must be modified to cope with larger logs and with the changes that occur in the wood of older trees.

Perhaps the most salutary lesson has arisen from the complete lack of tending of forests of the boom planting period, and therefore the inability to supply good quality logs to industry. This has restricted the uses to which radiata pine wood could be put. On the other hand, the fact that industry has become so well established on relatively poor logs can be turned to advantage. When good logs finally do come from forests being tended now and in the immediate past, the profitability both of growing and of manufacturing will be much greater.

Utilisation and the Development of Intensive Management

Thirty-year-old radiata pine trees in stands growing on the best sites have an average diameter of about 20 in. and on second-quality sites of about 17 in. Trees of these dimensions yield logs large enough for sawmilling; logs suitable for pulping are produced much earlier. The radiata pine plantings of the boom period therefore began to become ready for utilisation in the 1950s. At the same time some of the areas of Corsican pine, Douglas fir, and other species planted by the Afforestation Branch of the Lands and Survey Department were approaching 40 to 50 years of age, and the time had come for examining their use. Industry began in earnest to use exotic species and developed sawmilling, pulp and paper production, veneer production, and timber preservation, all on a comparatively large scale.

Not until this utilisation had developed were many of the problems of exotic forest management revealed. The need to log as economically as possible soon showed that the patchwork planting to several species, carried out in the first plantations to be established, added considerably to costs. Arguments about thinning and the intensity of thinning were largely theoretical until industry showed its ability to use the wood so obtained. Because the exotic forest products industry is young and because most stands of some important species such as Douglas fir are as yet only halfway through their rotation, knowledge about management is still incomplete. Information has been accumulated mainly about radiata pine, since most of the usable forest is of this species and it grows fast enough for the effects of different treatment to be assessed within a fairly short period.

Management of Radiata Pine

The tree occurs naturally in three small populations within 125 miles of one another on the coast of California and on nearby Guadaloupe Island. There is also a small population on Santa Cruz Island which is possibly radiata pine. In spite of its very limited natural distribution it is perhaps one of the most variable of pines. This variation can be seen in any planted area in New Zealand, and recognition of it and understanding of the reason for it are important to future management. Apart from variation of vegetative characters, such as cone shape and needles, characters important to management vary. Volume growth of main stems differs greatly; a few exceptional trees have been found whose increment far exceeds the average, and by selection it should be possible to increase the production of an already remarkably fast-growing species. Straightness of the main trunk also varies. Most trees have a tendency to lean or wave, but some trees in a stand grow perfectly straight. Branching also differs from tree to tree and at various ages in the same trees, though the variation in this character is not so important because of widespread adoption in recent years of pruning.

For the forester perhaps the most important properties of radiata pine are its ability to put on diameter growth rapidly—1 in. per year is common, 2 in. not uncommon—and the remarkable speed at which diameter growth increases after thinning. These characters make the tree one of the most amenable to thinning, and hence to manipulation of log sizes in stands. On sites where growth is average it will gain 300 cu.ft per acre, and on good sites will do much better. Thus it is possible to thin stands economically at frequent intervals; in fact the speed of growth makes it important to do so. The rapid diameter growth of radiata pine makes pruning profitable. If this operation is carried out at an early age, clear wood is grown once the pruned stubs have healed over. The additional value of this clear wood, in the lower logs of the stem at least, far exceeds the cost of pruning.

On most sites the tree regenerates readily, and on the pumice soils prolifically. Though the closed cones containing seeds remain on the tree for many years without opening in all but the hottest parts of New Zealand, when stands are clearfelled the closed cones rot on the ground and the seed in them soon germinates. The tree's ability to regenerate naturally was demonstrated after the Taupo fire, already mentioned, in which some 30,000 acres of 14-year-old radiata pine growing on pumice country were burned. The fire opened the cones, liberating seed, and the whole area regenerated quickly, in places to the extent of half a million seedlings per acre.

Radiata pine requires what might be termed average-to-good sites in New Zealand, although it will grow under a wide range of conditions. It thrives best and puts on the greatest volume on the pumice soils, but develops probably its best form in the Nelson region, where it has been planted extensively on the Moutere gravels. With the assistance of phosphate fertiliser, radiata pine will grow satisfactorily on a range of the stiff North Auckland clay soils, and it is almost the only tree grown on dunes in sand-fixation projects. It has been planted successfully from the North Cape to Stewart Island. Young trees will not stand out-of-season frosts because frost injury to the leaders, which grow almost throughout the year, predisposes them to attack by a fungus (*Phomopsis*).

In State forests all new plantings of radiata pine and regeneration after clearfelling of untended stands are thinned and pruned. Some afforestation companies are also introducing very intensive thinning and pruning programmes, and in the forestry encouragement loan scheme part of the loan is conditional on tending after establishment. There is little doubt that radiata pine as a timber tree will be greatly improved in the foreseeable future.

Early thinning out of malformed trees in stands leads to some improvement in tree form in later crops, but breeding and selection investigations are also being carried out to provide better trees. Breeding is being studied in a long-term investigation, but meantime the best formed trees that can be found are propagated and planted in "orchards" for the production of good seed.

Over about 40 years radiata pine has been developed from a forest tree with an uncertain future to the country's most important tree. It is being grown intensively, and as its characteristics come to be better understood, and breeding and selection have their effects, it should take its place among the world's most intensively managed forest trees. The growing of it in exotic forests has certainly become one of the main factors to be taken into account in any consideration of New Zealand forest policy.

Douglas Fir and Corsican Pine

Two other trees, the North American west coast Douglas fir or Oregon pine and the European Corsican pine, have to date shown qualities that set them out from the many other forest tree introductions.* They stand next in importance to radiata pine for growing south of the central North Island. Douglas fir needs the best sites that the New Zealand forester can acquire for planting; radiata pine grows on a wide range of average-to-good sites, and Corsican pine on sites too difficult for radiata pine. These three trees will grow on most sites at present available and have become the major exotic species planted. Others are used for special sites—mainly very difficult ones—or for planting north of the central North Island, where neither Douglas fir nor Corsican pine grow satisfactorily.

In the north, *Pinus caribaea*, *P. elliottii*, and *P. patula*, three species of pines from the southern areas of the United States of America and Mexico, are planted on a small scale, though in this region the intricate pattern of soils, on some of which none of these trees will grow, has caused the forester to look for improved types of the maritime pine, *P. pinaster*. This European, Atlantic, and Mediterranean tree thrives throughout the north, but types are very poor. It is said that at one time there were good types, but that in regeneration, following successive fires, they have been supplanted by poor forms. In certain areas of its natural distribution, notably in Leira in Portugal, there are good forms, and New Zealand and Australia have introduced them.

Douglas fir has a remarkably extensive natural distribution in western North America, growing throughout 2,000 miles of latitude and with an altitudinal range of 5,000 ft. It is one of the best known of the world's forest trees, mainly because of the excellence of its timber, which has gained markets far and wide. A strong, durable softwood available in large sizes, it is especially valuable for structural purposes. The first introductions in this country were fortunately of good types, as was all

---

* Recent attacks by *Dothistroma pini* may change the relative importance of Corsican pine.

seed imported subsequently. Since the tree seeds at an early age, locally produced seed has been available for over 30 years. Regeneration is prolific, even better under favourable conditions than that of radiata pine. Douglas fir grows on a fairly wide range of sites, but does best on moist, deep pumice soils. Volume growth is very high after about 20 to 30 years, reaching 500 cu.ft per acre mean annual increment[1] or more in the best stands. It seems that high volume growth can be maintained for a long time, because the oldest stands in New Zealand (about 100 years old) still put on good increments. (A 90-year-old stand at Mount Peel in South Canterbury is putting on total volume growth of up to 400 cu.ft per acre per year). Most stands are less than 50 years old, but thinnings from them yield supplies of pulpwood, good posts and poles, and some timber. Large structural sizes cannot be cut from the size of logs at present obtainable but they will be in time. On present indications Douglas fir is likely to become New Zealand's second most important exotic forest tree.

Corsican pine, *P. nigra*, the third most important species planted in State forests, has a wide distribution in southern Europe from Spain to the Crimea and the Caucasus. It has distinct geographical forms within its range, and one of the poorer of these, now known as *P. austriaca* or Austrian pine, was included, with good forms, in the first State plantations. Seed of poor forms was also brought in during the planting boom. Corsican pine can withstand more rigorous climatic conditions than can radiata pine, but does not put on such a high volume of growth. Its use could be restricted because of susceptibility to *Dothistroma*.

Other Exotic Forest Trees

A few of the other exotic forest trees planted extensively are already utilised on a minor scale. In time these will undoubtedly occupy major positions in planting and utilisation programmes. Looking still further into the future, the Forest Service is undertaking a vigorous programme of introduction of other trees, of provenances[2] and varieties of these, and of further seed of trees already introduced. A very wide investigation of forest trees of possible use is therefore assured.

[1] The total increment up to a given age divided by that age.
[2] The geographical source or place of origin from which a given lot of seed or plants was collected.

Perhaps the most remarkable introduction as far as adaptability to growing conditions and ability to regenerate are concerned is *Pinus contorta*, the lodgepole pine from western North America—the region that supplied our radiata pine and Douglas fir. It occurs from the sphagnum moss bogs of the Alaskan Coast to southern California and eastwards to the Cascade Mountains, a range of 30° of latitude and 11,000 ft of altitude. Within this range there are distinct geographical forms.

Lodgepole pine has come under special notice in recent years because regeneration from established plantations and shelterbelts has invaded unused land or very lightly stocked farmland. Most concern has been felt about regeneration coming mainly from the Karioi State Forest on the southern flanks of Mount Ruapehu. From plantings made there between 1925 and 1935 lodgepole pine spread, in varying densities, over tens of thousands of acres of tussock grassland and up to 6,000 ft, or well above the natural timber line on Mount Ruapehu. This infestation has now been mainly cleared.

Control of any forest tree that establishes freely over extensive areas where it is not wanted is sure to be costly, but there is no reason for condemning lodgepole pine, which has many virtues; the unwanted regeneration should have been checked early when its vigour became apparent.

Lodgepole pine tolerates an amazing range of sites and, so far, is the most promising tree for checking erosion at high altitudes, where often the soil is poor and the climate severe. It is keenly sought in its native home for pulpwood. If the pulp and paper industry expands markedly in New Zealand, it will probably be necessary to grow lodgepole pine extensively on harsher sites in the South Island to provide enough raw materials. There are several million acres where it could be grown and where it would probably yield much more than would present use of this land.

Next to radiata pine in extent of planting until recently was *Pinus ponderosa* or western yellow pine, another tree from western North America.* Because of its resistance to heavy frost damage it was used during the 1925-35 planting boom on sites where radiata pine would not grow. The natural range of the tree covers a huge area from British Columbia to northern Mexico and from the Pacific Coast to the dry western interior

---

* Recent attacks by *Dothistroma pini* may alter the relative importance of this tree.

of North America. Within this range are many forms of the tree and, unfortunately, we inadvertently introduced some of the poorest, which have produced small increments on sites that could have been occupied by much faster growing trees. These introductions, especially of the stunted form of *P.ponderosa* var. *scopulorum* are a clear warning of the danger of collecting optimistically from trees other than good, tested sources. By contrast New Zealand has some very good stands from seed of good origin.

European larch (*Larix decidua*) was planted over several thousand acres when afforestation by the State commenced. One of the largest concentrations of the tree, probably anywhere, including stands in its natural habitat, is the 5,000 acres in Whakarewarewa and Waiotapu Forests (a subdivision of Kaingaroa Forest) of the Rotorua district. Many of the stands there, although untended, are of good form and vigour and yield post and pole material, but the timber from saw logs has not found a ready market. When trees are older the timber might be accepted for specialised uses, but, meantime, small acreages only are planted. A related species, Japanese larch (*L. leptolepis*) is also planted on a small scale. It is notable for its fast growth, and it hybridises with European larch.

Monterey cypress or macrocarpa (*Cupressus macrocarpa*) has been one of the most widely planted trees on farms. It is still planted extensively but is subject to a canker disease (*Monochaetia unicornis*), which attacks many, if not most, trees to some extent and kills them when growing conditions are adverse. Its relative, Lawson's cypress (*Chamaecyparis lawsoniana*) was at one time planted even more generally for shelter and ornament, but canker killed it over extensive areas. Planting stock became infected in nurseries, and through it the disease was spread widely. The same disease killed macrocarpa when it was planted extensively throughout the highlands of East Africa.

Timber cut from macrocarpa logs from open-grown trees amounts to almost three million board feet (1966) and finds a ready market for joinery. However, because the tree is disease-prone it is not always safe to plant. Another cypress (*Cupressus lusitanica*), resists canker and may, in time, replace macrocarpa. It hybridises remarkably easily with macrocarpa and much of our planting stock consists of hybrids.

The Californian redwood (*Sequoia sempervirens*) has been planted over several thousand acres in New Zealand, but so far

A superbly grown plot of radiata pine on a farm in the Wairarapa. It is 35 years old and has been pruned carefully from an early age. It is an example of the quality of tree that can be grown.

(Page 65)

Above: On most sites radiata pine regenerates readily and on pumice soils prolifically. Fire in a 14-year-old stand extending over 30,000 acres was followed throughout the burnt area by regeneration similar to that illustrated.

Below: Radiata pine natural regeneration, thinned and low pruned. There is a heavy undergrowth, largely of native shrubs.

has succeeded on special and most favourable sites only. On these it grows well and rapidly. The best known stand is that, now reserved, in Whakarewarewa Forest, Rotorua. Western red cedar (*Thuya plicata*), too, has been planted over several thousand acres, also with limited and local success. The purpose was to produce joinery timber, but virtually nothing is yet known of the locally grown timber.

Among the unaccountable failures has been the North American western hemlock (*Tsuga heterophylla*), one of the main forest trees in north-west North America. In New Zealand the species has never reached even moderate tree size. Scots pine (*Pinus sylvestris*), the main timber tree of Europe, has seldom succeeded in New Zealand. The tree seems to be checked by mildly unfavourable growing conditions.

Introduced hardwoods have, by and large, failed as forest trees, because of inability to grow well as stands and because of absence of a market for the timber—or rather failure to produce enough of acceptable grades. Many species of eucalypts have been introduced and some have been planted extensively as farm and forest trees. A few have succeeded as timber trees on selected sites, but only now are we certain enough of them to grow them on a commercial scale.

Also introduced have been many species of poplar and willow, which are planted extensively for river control and soil conservation. The crack willow (*Salix fragilis*) has been used so extensively in river-control work and propagates so readily from twigs that much more money is spent in clearing river channels of unwanted willow trees than is spent on planting them for control work. This tree is to be seen on most riverbanks, sometimes in very large quantities, but its wood is hardly ever used. Poplars are grown on a very limited scale for the production of commercial wood, a limitation imposed by their need for deep fertile silts which are used for intensive livestock farming and cash cropping.

Working Plans and Sustained Yield

The ultimate expression of forest management is contained in what a forester terms a working plan, one formal definition of which is: "A written scheme of management aiming at continuity of policy and action and controlling the treatment of a forest."

The need for working plans arises mainly from the time

factor in forest management. Because of the long time trees take to grow, continuity of management is essential. Otherwise successive managers are likely to endeavour to impose their own preferences and tend to decry the methods of their predecessors. This would lead to radical changes during the life of a forest, and these can be damaging in the long run. Apart from laying down prescriptions for continuous management, working plans are convenient repositories for summaries of information about forests.

The first working plans known were compiled for some German forests about the beginning of the 1800s. They were written around the regulation of yield, which was then the most important aspect of management because of the widespread over-cutting of forests. A regular yield must be geared to the ability of a forest to grow, and give industry a steady flow of material and its owner a continuous return. In European and many other countries the compilation and continuous review of forest working plans is now a highly developed part of forestry practice.

In New Zealand the 1921 and 1949 Forests Acts prescribed working plans for State forests. The clauses in the later Act read:

(1) The Director shall from time to time cause to be prepared working plans for all State forest land.

(2) Every working plan to which this section applies shall, subject to the rights existing when the working plan comes into operation, regulate as hereinafter provided the management of the land described in the working plan for such period not exceeding twenty years as may be stated in the working plan and in conformity with the objects of management therein stated.

(3) Every such working plan shall specify with respect to the working plan period—
    (a) The silvicultural operations to be carried out; and
    (b) The maximum area from which forest produce may be disposed of or the maximum quantity of forest produce that may be disposed of or both, as the Director thinks fit; and
    (c) The protection and development operations to be carried out; and
    (d) Such other matters as the Director thinks fit.

(4) Every working plan shall be subject to the approval of the Minister, and, when so approved, shall have effect according to its tenor from a date specified therein, and shall not be altered save by the Minister on the recommendation of the Director.

Since that Act was passed, 5.6 million acres (1967) of State forest have been put under working plans. As is to be expected in a country where scientific forestry has lagged until comparatively recent times, the contents of these plans vary greatly. Those for some indigenous forests in which management is minimal contain general descriptions and broad management prescriptions only. Those for some exotic forests in which management is becoming intensive are very detailed. When the log sale (which ultimately went to Tasman Pulp and Paper Co. Ltd.) from Kaingaroa Forest was advertised, the working plan was made available and prospective purchasers were advised to study it.

In 1965 an amendment to the 1949 Forests Act provided for the creation of State Forest parks, which are State forests containing areas that are especially desirable for recreation. Before a working plan for any State Forest park is approved by the Minister of Forests it will be made available for inspection by the public or by the committee set up to advise the Minister about recreational aspects of forest park administration. Objections to any working plan proposals may be made to the Minister. This liberal formula for working plans is desirable for forests in which recreation uses will be a dominant feature. It might well be asked why these forests are not made national parks. This would prevent them from serving several purposes and being managed accordingly. Some forest parks contain protection forests showing signs of depletion and erosion caused by browsing animals. Some contain areas of forest which will be logged; others include areas from which urban water supplies are drawn.

Even before new legislation was enacted, the Tararua Forest, covering some quarter of a million acres of forest on the mountain ranges between Wellington and Palmerston North, was managed on the principles now laid down for State Forest parks. Emphasis was given to the development of recreational facilities, mostly by erecting or subsidising the erection of huts and forming tracks for the large numbers of trampers and hunters who use these mountains.

A few companies and local bodies have working plans for

their forests, the 1927 plan for forest controlled by the Wellington Water Board being the oldest in the country. The board had 86,000 acres of indigenous State forest at the southern end of the Tararuas vested in it by the Wellington City and Suburban-Water-supply Act. A forestry section (Part V) of the Act gave the board control and management of the land coming under it and prescribed that: "The City Council shall from time to time prepare working plans in respect of the whole or any part of the area which it has appropriated for forestry purposes." The working plans were to be subject to the approval of the Commissioner (now Minister) of State Forests.

The regulation of yield has been one of the main purposes of working plans since they were evolved in Germany over 150 years ago. In many parts of Europe the destruction of forest close to areas of increasing population before the days of easy transport of a bulky commodity like wood brought about severe local wood shortages. This led to the close control of cutting and eventually, when forests were improved, to their management on a sustained-yield basis—that is, a constant and perpetual yield harvested as are annual crops from farms. A forest puts on a certain amount of new wood growth each year and if the forest is well managed, this growth can be cut continuously without detrimental effect. Skilled management can often increase it.

6

FOREST PRODUCTS INDUSTRIES

Over about 150 years our wood-based industries have progressed from the pitsawing of native kauri to the operation of large and fast newsprint-manufacturing machines using wood from planted forests of introduced species. Moreover, the stage has been reached at which the State has developed a policy of increasing afforestation to provide yields of wood which will support a continuing expansion of forest products industries. These are already important domestically and as exporters and their importance will increase considerably. The value of annual output of the whole industry had reached $254 million by 1966. Increased afforestation will lose much of its point unless the wood grown can be bought and used by thriving and enterprising industries.

Some doubting Thomases contend that wood as a raw material is being or is likely to be supplanted. Substitutes for the traditional uses of wood loom up from all directions, the plastics millenium is upon us, so why grow wood? Quite the reverse trend is shown by the consumption of wood in its various forms in the most highly industrialised countries. Wood is one of the cheapest renewable raw materials in the world, and although it might be losing some ground to substitutes in the construction field, rapidly increasing amounts are needed for making pulp and paper and building boards. Moreover, the chemistry of wood is only half-revealed; when it is fully known there is bound to be a great leap forward in its use as a raw material for manufacturing, including plastics.

Sawmilling Industry

The first sawmills started about 1840. The industry quickly grew to the stage of supplying most of the needs of the developing country and, for a time, of providing a substantial part of its cut of kauri and kahikatea for export. For almost 100 years it drew almost all logs from native forest, which receded rapidly as farming developed. Sawmilling was a transient industry in many places, fostering pioneer settlement by providing employment in the "backblocks" and helping to clear the "bush" for farming. Eventually as supplies of logs ran out in one district after another only great heaps of sawdust on partly developed farmlands were left as evidence of the pioneer days.

Very little native timber is now milled between the central North Island's King Country, where activity was greatest, and the shores of Cook Strait at the extremity of the North Island; the great kauri milling industry of the north has vanished. The last sizeable strongholds of indigenous timber cutting are on the west coast of the South Island (where the rimu forests grow on soils that cannot be farmed) and in the centre of the North Island (where agriculture developed late on the pumice soils and forest clearance therefore lagged behind that in other parts). Small sawmilling centres handling native timber remain in a few other districts, the main one being in western Southland, where operations are sustained partly by southern silver beech.

The first mills had circular saws driven by steam engines. The general type of equipment did not alter greatly over the years, though deal frames were added in the large mills built for the cutting of kauri. Circular saws are still widely used, but steam power has been replaced by diesel or electric motors, which means that off-cuts once used for firing are now burned as waste. The standard of "housekeeping" of most of these mills is very low indeed and often the timber is not cut accurately to the prescribed dimensions. The timber itself is more often mishandled than not, but deterioration is not common because of the inherently good properties of the wood of native species.

At first logs were extracted mainly by bullock teams, though in the kauri forests floating was practised where the stands were heavy enough to get together a large accumulation of logs. Later, the almost universal method of extraction was by steam winch and haulage of logs from the winch site to mill by steam-locomotive drawn tram. Inevitably felled forest was set alight from time to time by sparks either from winch or locomotive, and consequently young growth of commercial trees left from logging was soon destroyed. When farming followed, this was immaterial, but if it did not, any prospects of a succeeding crop soon vanished, leaving desolation.

While the indigenous forest sawmilling industry was still flourishing in most parts of the country, Canterbury had cut its small area of native forest and milling turned (about 1902) to radiata pine. Mills of the type that cut indigenous timber were used, and in the dry Canterbury climate they cut radiata pine logs satisfactorily.

By 1935 some of the first-planted State forests were reaching an age at which material was becoming available in large quantities in concentrated areas. The largest amounts were in the radiata and Corsican pine and Douglas fir stands in Whakarewarewa Forest and older parts of Kaingaroa Forest. The greatest potential, of course, was in Kaingaroa and in the radiata pine being planted by companies in the nearby district. The next largest quantities, also of the same species, were in the Tapanui group of forests on the border of Otago and Southland. All these forests were virtually untended from time of establishment but whereas large areas of radiata pine stands had grown beyond tending, Corsican pine and Douglas fir stands were only part way to maturity. The only material available from them would be in the form of thinnings.

Because the sawmilling industry's experience was almost entirely in the cutting and marketing of indigenous timbers, no interest was displayed in large-scale milling of exotic trees. Outside Canterbury, the timbers were foreign to the industry apart from the case-making trade; logs were much smaller than those from indigenous bush, and sawmill equipment quite different from that of circular-saw mills was needed for efficient sawing.

The Forest Service therefore entered commercial logging and milling in plants built at Waipa, Rotorua, in 1939, and at Conical Hill, adjacent to the Tapanui forest, where a pilot plant began in late 1948. Though these units had to be fairly big to operate commercially, they were also to serve as demonstration and development units both for production and for marketing techniques for the sale of exotic timber.

This pioneering effort proved to be essential for the impending transition from high grade, largely defect-free indigenous timbers, that were known and understood throughout the building, joinery, and furniture and cabinetmaking industries, to markedly defective timbers sawn from untended exotic forest logs. The operation of the two State mills laid the basis for the use of exotic sawn timber, the cut of which increased from 37 million bd.ft (11% of total rough sawn timber cut) in 1938, the year before Waipa began cutting, to 500 million bd.ft in 1966/67 (67% of total rough sawn timber cut).

The change was not only from high quality to low quality timbers, but away from timbers which even if roughly handled in the yard or on the building site gave a good end-result in finished timber or in building. The heart grades of native timbers were durable without preservation and very little kiln seasoning of them was done. Radiata pine on the other hand was a defective timber requiring careful grading. It sapstained rapidly and decayed quickly when exposed to moisture. Research was necessary to formulate grades and to determine seasoning and preservation practices and strengths and often physical properties of the timber. All this knowledge was applied through a large commercial unit. It was soon demonstrated that radiata pine could take the place of indigenous timbers for many purposes; for some uses it could not, but for others it was better.

The way was paved for industry to use sound techniques of sawing and handling radiata pine timber when the time came to begin utilisation of the plantings of the 1925–35 boom. Several big mills started up, two larger than either of the State mills, with the result that the sawmilling of exotic timbers differs considerably from the milling of indigenous timbers. Most of it is sawn in a few large mills using log sorting and handling and frame saws. The timber is dipped to avoid sapstain, a percentage is kiln dried, and almost half of it is treated with preservatives. (New Zealand has by far the largest use of preserved timber per head of population of any country.) The now freely available treated radiata pine is more durable than the heart native timbers once insisted on for much of our building.

Almost all the exotic timber handled by sawmills is radiata pine. So far, Corsican pine timber has been cut from thinnings only, mainly in the two State mills. Though it has some individual characteristics, it is usually bracketed with radiata pine for marketing. Douglas fir, also sawn from thinnings, provides 3.5 per cent (27 million bd.ft) of the total sawn timber cut. An export trade in radiata pine and some Douglas fir timber has grown up with Australia. So far this has not exceeded 45 million bd.ft in any year because of the low grades available.

Exotic timbers can now adequately fill requirements for general building and construction. They cannot, in the grades at present available, fill the requirements of house weatherboarding (except in finger-jointed form), high grade flooring and joinery, and high-quality furniture. These needs still must be met by indigenous or imported timbers. Defective exotic timbers are improved on a limited scale by finger-jointing and edge-bonding, and structural members are made up by lamination, but the physical properties do not match those of the indigenous

Douglas fir, Mount Peel, Canterbury, close to 90 years old. This stand is still putting on 300 to 400 cu.ft per acre per year.

(Page 68)

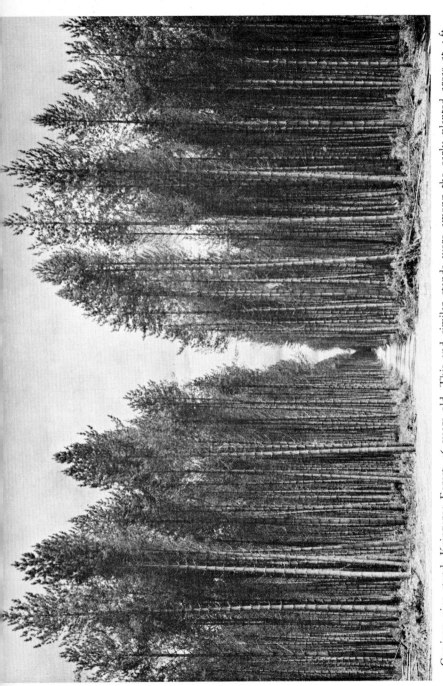

Corsican pine stand, Kaingaroa Forest, 46 years old. This and similar stands were among the early planted areas at 4 ft by 4 ft spacing. Trees are about 90 ft high. They are being thinned in spite of having been left untended. Yields of wood are very high—8,000 cu ft per acre to a 4 inch top for the stand shown.

(Page 67)

Lodgepole pine (*Pinus contorta*) stand, 40 years old, Hanmer Forest. In the foreground is shown prolific regeneration. This pine is perhaps the most remarkable introduction as far as adaptability to growing conditions and ability to regenerate are concerned. It came from western North America, the home of radiata pine and Douglas fir.

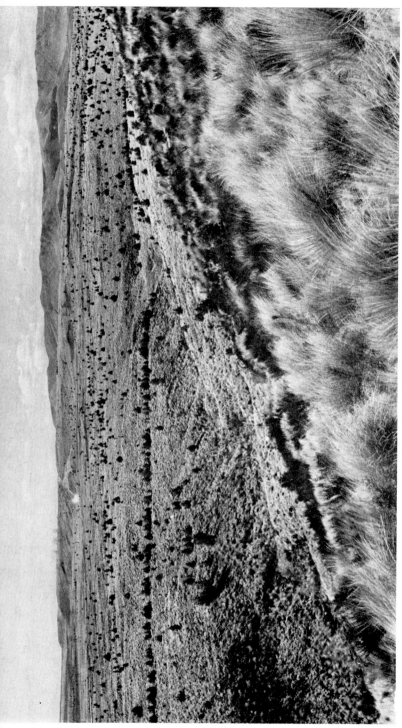

Lodgepole pine regeneration spreading from plantations on the slopes of Mt. Ruapehu through natural tussock grasslands. This regeneration seeds when it is very young. Eventually, the invaded grassland would be occupied by a block of trees.

(*Page 69*)

or selected imported timbers. Research will undoubtedly help to extend the uses of exotic timbers, and the production of clean grades will help still further, but some special needs will never be entirely filled by the exotic timbers we grow now. It is therefore important to perpetuate native timbers wherever possible and examine critically the range of exotic species we can grow with a view to selecting those which might satisfy our needs.

Plywood Industry

Indigenous logs were first peeled for veneer in 1911. Today the plywood industry uses principally logs of indigenous rimu, kahikatea, matai, kauri, and some minor species when obtainable. Most of New Zealand's needs are now supplied, except for special plywoods, which are imported: some veneer logs are imported for making decorative or marine plywoods locally. The supply of indigenous peeler logs is now dropping rapidly in the North Island as forests are cut out and the only substantial source is on the west coast of the South Island.

In spite of the poor quality logs available, radiata pine is rapidly filling the gap. Logs are proving exceptionally easy to peel. Defect-free veneer can be produced from lengths between branches that are long enough, but large quantities of radiata will not be peeled until logs that have been pruned are available. In time, the production of veneers and plywood from radiata pine could be an important industry, and the growing of logs for it is part of the policy of management of State forest and of some companies. The indications are that our Douglas fir forests, even though much faster growing than the natural forests of North America which produce good peeler logs (on which the large American veneer and plywood industry was founded), will yield logs capable of producing veneers. There has been a recent rapid growth of a section of the American industry based on average quality southern pine logs. There could be similar development with radiata pine in this country. It has been claimed that the plywood industry as a whole is the fastest growing industry in the United States.

Use of Roundwood

Readily available supplies of wood in the round or split for posts, poles, and piles are important in a rapidly developing country. Settlers carving farms out of the New Zealand bush obtained all the fence posts and rough building material they needed from this forest. Totara, because of its durability and ease of splitting, quickly became widely known, and the renowned durability of silver pine made it sought after far and wide. As native forest areas were stripped of the good fencing timber many farmers planted exotic trees, such as eucalypts and macrocarpa, from which they could produce their own supplies. However, as this did not prevent shortages, an import trade in Australian-hewn eucalypt posts grew up and sometimes Malayan hardwood fencing material was also brought in.

It was soon found that exotic forest produce, if treated with preservatives, made excellent posts. At the beginning of this development European larch and Douglas fir thinnings were used and treated in a pressure plant at the State's Waipa mill. When the difficulties of seasoning radiata pine in the round were overcome posts of this species were treated with multisalt preservatives by pressure. Very large quantities of radiata pine posts are now marketed (more than 4 million in the year ending 31 January 1967). The market for posts gives the forest owner some profit for thinning.

The first produce obtained by Europeans from New Zealand's forests was kauri for ships' spars, but very few poles of this tree have ever been produced, though kauri and totara have been used extensively for wharf piles. Imported Australian hardwoods (eucalypts) became the traditional pole timber, and Australian turpentine (*Syncarpia laurifolia*) the piling timber. Australian hardwood was also the main material used for railway sleepers. Imports of these three commodities rose to quite substantial figures, but as with posts, exotic woods soon entered the market. Larch and Douglas fir, pressure treated, were first in the field for pole supplies also, but more recently, radiata pine poles have been produced in increasing quantities. These three species and, to a less extent, Corsican pine, have replaced all imports of smaller poles, but so far they have made little impression on markets for large poles (used mainly for electric transmission lines) or piles. No doubt, further research will assist such replacement. Within the past few years, treated radiata pine sleepers have, almost drastically, largely replaced imported eucalypt sleepers.

Export of Logs

Within the last decade a log export trade with Japan has developed. The logs are mainly radiata pine sawlogs from

1925–35 planting mainly in the central North Island forests, although smaller quantities have from time to time been exported from the South Island. The logs are at present surplus to the needs of New Zealand industry. The amount exported has reached a substantial volume—approaching 30 million cu.ft in 1967, which was about fifty per cent of all the wood—in various forms—exported.

The export log trade has been conducted mainly through the port of Mt Maunganui, in the Bay of Plenty, which was built especially for the export of forest products, mainly pulp and paper. The shipping of logs was not visualised at the time the port was constructed, and to cater for it wharves had to be extended. Revenue from the traffic has helped substantially to build up harbour facilities. With the recent extension of the trade to other ports harbour authorities are aware of the financial benefits this type of trade can bring.

Production of Paperboards and Paper

New Zealand's consumption of paperboards, wrapping papers and newsprint had reached about 40,000 tons by the early 1930s. At this time, pulp, kraft paper and newsprint had already been produced experimentally from New Zealand-grown radiata pine at forest products laboratories in the United States and in Australia. One of the afforestation companies, Timberlands Woodpulp Ltd, later renamed Whakatane Board Mills Ltd, considered that the time was ripe to produce paperboards, even though its own trees were not 10 years old. It put forward a proposition for obtaining radiata pine thinnings from the older-established State forests in the district. This was refused, however, because the Forest Service had in mind the production of newsprint, for which large supplies of raw material would be needed. Undeterred, Timberlands Woodpulp bought plant capable of making a large range of board products from varying mixtures of imported chemical pulps and groundwood produced from radiata pine. After many difficulties had been overcome the mill was built on the Whakatane River close to the town of Whakatane and production began in July, 1939. Some of the pulp was produced from 11-year-old radiata pine thinnings obtained from the plantations of another afforestation company on Matakana Island at the entrance to Tauranga Harbour. Although good products were produced there was considerable opposition to the sale of them from importers of similar materials.

The outbreak of war in 1939 changed the position almost overnight, as the company was able to produce sorely needed articles that could be imported only with great difficulty and then at high prices. A difficult trading position changed quickly to a satisfactory one, although prices were controlled and were lower than world prices. As time went on the company operated under increasingly favourable conditions. The spectacular rise in overseas prices in the late-war and early post-war period put an end to the threat from imported boards. The growing consciousness of the value of economic local industries in conserving overseas exchange also aided the company's development. The decision to pack butter in cartons rather than in wooden boxes, partly because of the increasing scarcity of kahikatea timber, created a new and large market. This made possible long production runs on the board machine and paved the way in the early 1950s for the purchase of a second machine and equpiment to produce long-fibred pulp by a semi-chemical process.

The history of Whakatane Board Mills has been given in some detail because of its importance as marking the first commercial pulping of radiata pine in New Zealand and the production of board from it. It was a simple but important beginning and the experience the company gained must have helped the next company—New Zealand Forest Products Ltd—to produce pulp and paper from radiata pine, in this instance kraft pulp and paper to supply mainly the New Zealand market.

New Zealand Forest Products Ltd had its origin as New Zealand Perpetual Forests Ltd, an afforestation company promoted in 1922 by a firm of brokers. The company sold bonds, initially for £25 ($50). For each payment it undertook to buy an acre of land, plant it and maintain the planted area for 20 years. In an energetic forestry-promotion and bond-selling campaign 170,000 bonds were taken up, in New Zealand, Australia and further afield. Land to match this number of bonds was bought and planted.

As the basis for an afforestation company, the bond structure had two main defects: the individual acre that a bond purchased and paid for planting could not be identified, and bondholders were not organised to sell their timber or to develop utilisation of it. The bond-selling company's contract was to plant and maintain the trees for 20 years, and there its responsibility ended.

Bondholders in many other similar companies were in an equally unsatisfactory position, which was one of the reasons for the Government in 1934 appointing a commission of inquiry into company promotion. This judicial body recommended the abolition of the bond system and the incorporation of bond money into share money in realisation companies. The Companies' Bondholders Incorporation Act 1934–35 was passed to permit the change. In 1935, therefore, bondholders of New Zealand Perpetual Forests Ltd were incorporated into New Zealand Forest Products Ltd, opening the way for the new company to take over the forests planted on behalf of the bondholders and to formulate utilisation plans. After protracted and difficult negotiations the new company found itself the possessor of 170,000 acres of growing trees (mostly radiata pine) varying in age from 1-13 years and of $2,694,000 in cash investments and various assets. Valuable assistance was obtained from the Australian Council of Scientific and Industrial Research, no doubt called on partly in an endeavour to protect Australian capital; the head of the Forest Products Laboratories of that council was soon engaged to report on utilisation plans.

The first steps were the building of an insulating-board mill, sawmill and case mill at Penrose on the outskirts of Auckland, and a sawmill in the forest at the centre the company named Kinleith. The Auckland plants were opened at the end of 1941, when the oldest trees were barely 20 years old.

In 1953 a kraft pulp mill of 25,000 tons capacity came into operation at Kinleith. Part of the pulp was to be supplied to the Australian Newsprint Mills at Boyer in Tasmania, and to use some of the remainder it was decided to build a kraft paper mill alongside the pulp mill. These were the first steps in the production of kraft pulp and paper in New Zealand. Kinleith, the mill site, and Tokoroa, the small centre where the company built a township for its employees, are now household words in the industry.

From this stage the company continually expanded its production of pulp and paper, sawn timber, building boards and other commodities. In 1961 it amalgamated with Whakatane Board Mills. Production of the two plants in 1967/68 ran at 85 million board feet of sawn timber, 200,000 tons of pulp, 160,000 tons of paper and paperboard, and 97 million sq.ft of fibre board plus other commodities. Including log exports the companies used about 55 million cubic feet of roundwood (about 30 per cent of the total exotic roundwood production in New Zealand). The company expects additional roundwood to be available from existing forests up to about 1980 or sooner, by which time the sustained yield will have been reached. Further expansion will then depend on new planting.

Thus an afforestation company has grown from unsure beginnings and obscure plans for development to be the largest of any sort in New Zealand. It is based primarily on home markets, but it also contributes substantially to exports. The original promoters struck the lucky combination of pumice soils and radiata pine, and although, until utilisation was well developed, the minimum of forest management was practised, it is now becoming very intensive.

The possibilities of a pulp and paper industry were in mind when the large acreage of exotic forest was being planted. The great advantages of the Kaingaroa Plains were seen: a large, mainly flat area where a big volume of radiata pine could be grown within short distances of suitable mill sites, water supplies and potential ports. There could have been few more favoured localities outside the virgin forests of North America. Even there the potential for sustained yields could not be nearly as great as in New Zealand.

When thoughts turned to the utilisation of this rapidly growing forest, planning began of a pulp and newsprint enterprise, combined with sawmilling. The economics had to be such that a large part of its output could be marketed overseas at world-parity prices. By New Zealand standards therefore it had to be a large enterprise.

In 1952 the Tasman Pulp and Paper Co. Ltd was granted a sale of 23 million cu.ft of wood from Kaingaroa Forest (equal to about one fifth of total roundwood use in the country at that time), with rights over an additional 5 million cu.ft per annum. The floating of such an enterprise was an exercise for New Zealand in the need for detailed, long-term planning, and the interlocking of Government policies and Government investment with investment by private enterprise and overseas companies which controlled substantial newsprint markets and could produce the essential technical management. A single newsprint machine with a rated capacity of 80,000 tons per annum was installed together with a sawmill having a capacity of 75 million bd.ft per year. There was to be a surplus of kraft pulp available for export. The first pulp was produced in July 1955 and, the first newsprint in October of that year. A second newsprint machine was installed in 1963. The output of the two machines has risen (1968) to close to 200,000 tons per year.

Tasman has been able to supply virtually all internal requirements of newsprint and to export a large proportion of its output to Australia, despite the fact that it came into production at a time when the world capacity for newsprint production was being expanded considerably and far outstripped world demand. This capacity will, in the next year or so, exceed the world demand of about 14 million tons by some 3 to 4 million tons. Tasman has therefore not achieved its success easily, because international prices of newsprint have fallen since it first came into production. Under these conditions the company has taken somewhat longer than planned to settle down to a sound economic basis.

The sale to Tasman of the large quantity of wood from Kaingaroa Forest (which has now risen to 60 million cu.ft per annum) has helped the Forest Service to begin intensive forest management. Age-class distribution is being improved, and mistakes in the original siting of species will in time be rectified. The forest, as predicted, is proving to be one of the most productive in the world compared with any area of similar size. It grows a large quantity of wood on a concentrated area and evenutally, under intensive management, the yield of this wood will be very much higher than it now is. Kaingaroa therefore forms an advantageous base for the inauguration of a pulp and paper industry geared for export.

Development of the Kaingaroa Forest and of the Tasman Company have transformed the Bay of Plenty district, because the Government has not only had to participate as a shareholder but to build port facilities, roads and railways, and erect houses and power lines. The rate of economic development in the district has been much higher than in any other part of New Zealand.

The lesson from this is that the development of pulp and paper units of this magnitude does not mean simply the growing of a forest and the building of a mill: it involves the country, nationally, in many ancillary developments.

7

LOOKING AHEAD

The long established and consistent policy of replacing decreasing indigenous wood supplies with exotic wood has now reached a stage at which the main needs of the country will be assured provided we grow successfully the necessary area of exotic forest trees. Thus the early planners' proposal to replace wood with wood, as native forests receded to make way for settlement and agriculture, has become fact.

Though we have had remarkable success in acclimatisation and use of exotic forest trees, foresters must, because of the long time-element involved, be cautious in their approach both to liquidation of remaining indigenous resources and to unreserved acceptance of exotic forests. There is a constant threat to these from introduced pests or diseases. Good though our quarantine services are it would be strange indeed if potentially dangerous fungi and insects did not come into the country occasionally, or if some part of those intercepted at ports did not escape the quarantine net. However, the skills of managing exotic forests are being improved rapidly. Intensive management will be possible in the long run because of high yields, and hence the ability to grow concentrated quantities of wood per unit area. And the more intensive practice becomes the greater should be the effectiveness of measures against epidemics when they come.

Success has been achieved not only in producing enough timber but also in producing high grades of certain types of pulp and paper from radiata pine. Governments have therefore supported increased planting to supply forest products both for home and export markets.

Government policy for the timber-yielding native forests is continuing conservation or rationing of the remaining supplies and, wherever possible, introduction of permanent management. The need for conservation is greater today than ever because of the inability as yet to obtain high grades of timber from exotic forests. Timber from these high grades, for which there are long-established and important uses, must be obtained from

The first sawmills started about 1840. From that beginning the industry grew to the stage of supplying most of the needs of the developing country and for a time of providing a substantial part of its cut of kauri and kahikatea timber for export. This is one of the earliest mills at Kaiwarrawarra, Wellington, driven by a water wheel.

(Page 75)

In kauri forests floating was practised where the stands were heavy enough to get together a large accumulation of logs.

native trees or imported. As supplies are becoming more difficult to obtain from both sources, it is important that the maximum quantity of high grades should be cut from the available native logs.

Domestic Needs of Forest Products

By comparing present with past consumption of forest products and taking into account the country's general and industrial development, it is possible to forecast long-term needs. Methods of making such prognostications are becoming refined, so that estimates should be reasonably accurate. Added reliability can be obtained by making comparisons with consumption patterns in other countries, especially the United States. Some trends in New Zealand resemble those in the United States, which has large wood supplies and a large consumption of forest products.

The most important of the many factors determining the level of use of forest products is of course the availability of raw material. If we did not have this in New Zealand, but indulged in our present use of forest products, the bill for imported timber, pulp and paper would be about $200,000,000 (1968). As we could not commit anything like this from overseas earnings, it would be necessary to look for substitutes for timber.

The effect of dependence on overseas exchange to buy forest products is to be seen clearly in the use of papers, particularly newsprint. The first newsprint was made in 1956; before then it was all imported, and not all of these imports have yet been eliminated. There has been a rapid reduction, however, so that in 1967 imports amounted to some 1,000 tons out of over 70,000 tons used. Consumption per head rose from 37 lb per year in 1950 to 69 lb in 1966 and is expected to reach about 80 lb by 1975.

The use of industrial papers (wrapping papers, cardboards etc.) shows an even more striking increase. In 1950 per capita consumption was 44 lb. In 1966 it has risen to 127 lb. By the year 2000 it is expected to rise to over 300 lb. The use of printing and writing papers has not shown such a marked increase because New Zealand still depends to a considerable extent on imports, which are controlled at a level which is almost certainly well below total demand. The per capita use rose from 18 lb to 24 lb between 1950 and 1966.

Over the same period, the per capita use of all papers more than doubled from 91 lb to 220 lb. The amount is expected to rise to between 400 and 500 lb by the end of the century and this per capita use approximately equals present United States use (479 lb in 1965).

In contrast to the use of paper, the per capita consumption of sawn timber is expected to fall from its present exceptionally high level of 279 bd.ft (1967). This is possibly the highest use in the world. Any fall in consumption, however, will be offset to some extent by marked rises in the use of wood panels and plywood, and in fact the use of these will probably more than offset any reductions in the use of sawn timber. Nevertheless, New Zealand is likely to remain a very large user of timber for a long time.

The projected rate of growth of use of all forest products at 2.1 per cent is predicted to outstrip the population growth of 1.8 per cent. A summary of total wood requirements for the home market and exports in the periods 1976-80 and 1996-2000 follows.

Supplies of Wood and Other Fibres for Domestic Consumption and for Exports

(Million cu.ft roundwood equivalent)

	Year ended March 1968	1976-80	1996-2000
Exotic wood	185	280	404
Indigenous wood	37	46	28
Imports	12	16	25
Sawmill and forest residues	19	35	47
Waste paper	3	5	12
Total	256	382	516
Domestic market	184	232	346
Exports	72	150	170

Potential for Exports

The afforestation target set by the Government in 1961 was to produce sufficient wood to provide domestic needs of forest products and yield a surplus of the order of 150 million cu.ft by the end of the century. This surplus would be converted to forest products for export. However, it is now evident that this amount of wood will be available by 1980. Exports in 1968 (June ended year) were equivalent to 87 million cu. ft. and their value was $41 million. Exports of 150 million cu. ft. in 1980 would probably have a value of roughly $100 million, though this will vary according to the kinds of produce sold. It could also vary considerably according to the amount of wood grown. On both sides—the kinds of produce and the amount of wood available to produce them—the export potential for New Zealand should be closely examined. Some of the main trends to be taken into consideration in such an examination are touched on in the following sections.

The most important fact is that Australia looms as a large market for New Zealand forest produce, taking a high percentage of the $41 million worth now exported. She is likely to remain the dominant market for some time. The development of the forest products trade with Australia suits the general trade pattern of both countries. New Zealand is the largest single market for Australian manufactured goods. In spite of the fact that New Zealand trades on a multilateral basis, the extent of this market must depend partly on reciprocal trade. This has been allowed to get sadly out of balance, favouring Australia in the ratio of approximately 4:1. The strongest possibility of New Zealand's being able to rectify the position lies in exports of forest products, and certain of these were the most important items included in the limited free trade agreement signed between the two countries in 1966. Under this agreement New Zealand should be able to expand forest products exports to Australia substantially. Because of the dominating importance of the Australian market, a brief account is given of the country's forestry and forest products potential.

Australian Potential

Though Australia is 29 times the size of New Zealand, an area of only four to five times its size receives enough rain to support the growth of forest. Within this area is concentrated all intensively farmed land and almost the entire population of 11.5 million. What forest resources remain have not all been assessed, but a figure frequently quoted is 25 million acres of permanently reserved forest capable of perpetual yield. The average yield quoted by Australian foresters is, however, only 25 cu.ft per acre. This surprisingly low figure is an average for the whole area.

In most Australian forests the main forest trees are species of *Eucalyptus*, of which there are a great number—some 600 species and many groups of hybrids—usually present as mixtures. Such mixed forests are not easy to work. In addition to the eucalypt forests there are smaller areas of "rain forest" containing mixed hardwoods and areas of cypress pine (*Callitris*) forest. Most of the mixed hardwood forest is in Queensland but there are some southern beech forests, mainly in Tasmania.

Thus nearly all Australian forests are hardwoods. Many of the eucalypts produce timbers that are difficult to cut and process, and the multiplicity of species complicate marketing. Nor do they produce good pulps, though they are used for this purpose. That such extensive use is made of, for the most part, difficult woods speaks volumes for Australian forest products research and the ability of the industry.

Like New Zealand, Australia under European settlement soon developed a timber economy for house building, and those parts of the country that were more or less treeless began afforestation with the same exotic trees as were introduced into this country. South Australia, which had very poor and meagre natural forest resources, set up a Woods and Forests Department in 1876—about the time New Zealand passed her first and abortive Forest Act—and began planting principally radiata pine on the sands in the Mount Gambier region. The State now has 150,000 acres of radiata pine plantations. Victoria and New South Wales also plant radiata pine. In addition New South Wales plants some U.S. southern pines towards its northern boundary, where the climate is unsuitable for radiata pine. Queensland plants these southern pines and small areas of its indigenous kauri (*Agathis robusta*) and hoop pine (*Araucaria cunninghamii*). Western Australia plants mainly maritime pine (*Pinus pinaster*) and Tasmania small areas of radiata pine and other species. Throughout the whole country over 650,000 acres (1967) of plantations have been established. This is about half the New Zealand total.

Australian forest products consumption and requirements are high, a reflection of a high standard of living. Her own forests, including both the natural and planted ones, cannot supply needs and her import bill had already risen to about $160,000,000 in 1964. An Australian forester[1] wrote recently:

"It does not require any great prescience to propose with confidence the inevitability of a 'grave' timber shortage in a lesser time than the seedling of today will take before it can make its future contribution to supply. There is no discernible fall-off in the upward trend of per caput consumption of

---

[1] McGrath, K. P. "*Approaches to the solution of Future Wood Supply*" Australian Forestry. 29:3 1965

wood as yet; and no reason to expect that, price being right, our population, which is expected to increase threefold in scarcely more than half a eucalypt rotation from now, will be looking for much less than three times the volume with which our present population is provided. At the same time we continue to find the great bulk of our requirement from capital—the age-old native forest bank from which we have made continued withdrawals and which, supplemented by an additional 20-30% imported, has met our requirements ever since Governor Philip proclaimed the original colony 177 years ago."

At present consumption levels and prices Australia's bill for imported forest products could be as high as $500 million by the end of the century. In the past few years the Commonwealth Government has begun to take serious notice of this situation. As a first step to coordinate forest policy an Australian Forestry Council has been formed. This council proposes a planting programme of 75,000 acres per year—twice the present rate of planting—for the whole country. However, the effects of this programme will not be felt fully until the end of the century and even then the wood produced could fall far short of requirements. There should, therefore, be ample opportunities for New Zealand exports.

Asian Markets

As the standard of living of the millions in Asian countries rises, so does consumption of forest products, particularly of paper, which is rising very fast. Canada, with its large wood resources and forest products industries, is probing some of these new outlets, which apart from Japan, are not yet large. New Zealand at present finds these markets difficult to penetrate, but later they must offer good opportunities.

Japan

Demands on Japanese forests by her forest products industries have increased greatly in the last decade. The supply of wood from them has doubled and increasing quantities of logs are being imported for sawmilling, pulping, plywood production and other uses. In 1968 Japan imported about 40% roundwood requirements. In spite of intensification of reafforestation, Japan cannot hope to bridge the gap between demand and supply from domestic resources.

In 1925 (the earliest year for which a figure is available) wood pulp production in Japan was 400,000 long tons. This rose to over 1 million tons in the fifties and now exceeds 6 million tons. Most of the available local raw material for pulp production is hardwood. Import figures show that an increasing proportion of pulp imports are of softwood pulps, indicating the growing need for long-fibre pulps.

Japan now has a population very close to 100 million and consumption of paper rose from about 60 lb per head in 1956 to 163 lb in 1964, when she assumed third place in world ratings of paper production. These phenomenal increases have not been matched by any other country, and demand on this scale has presented New Zealand with exceedingly promising marketing opportunities. In time Japan is likely to become the most important market for our forest products.

Philippines

The population of the Philippines is 30 million and consumption of paper rose from 8 lb per head in 1956 to 14 lb in 1964.

Indonesia

This country, with a population of over 100 million, has a very low consumption of paper and board—under 2 lb per head. Very small quantities of pulp are imported from Sweden, Japan, and USA for manufacturing.

Mainland China

China, with a population of some 700 million, has an estimated consumption of paper and paper board of 9 lb per head, which is treble the 1956 figure. In 1964 pulp imports were 35,000 short tons and paper board imports 30,000 short tons. Pulp was imported principally from Scandinavia, and paper and paperboard from Japan, Finland, Sweden, and Hong Kong.

Taiwan

Paper and paper board consumption reached nearly 50 lb per head in 1964. The population is just over 12 million.

Thailand

A population of almost 30 million had a per capita consumption of paper and board of 7.7 lb in 1964. This was more than three times the 1956 figure.

The World Scene

Modern methods of forest survey and inventory have revealed that, far from facing a wood famine, as was predicted at the end of the First World War, the world has very large areas of unexploited forest. The role that these will play in providing wood for expanding demands for forest products is still largely unknown but is being examined more and more closely. One certain thing is that these forests can be tapped only if the wood supplies so obtained can be used economically. The cost of opening them up and of the transport, over long distances, of wood or manufactured forest products are the usual problems.

A distinction must be made between softwood and hardwood resources. The former now provide probably well over three-quarters of all the wood used in the world. The main industrial use of hardwoods is for furniture timber and veneers, although pulping of them is expanding, and very large quantities are used for firewood.

There are extensive temperate zone hardwood forests throughout the Northern Hemisphere, particularly in Europe, where they are fully utilised, and in North America, where they are only partly so. Much more extensive are the tropical hardwood forests in South-east Asia, Latin America and parts of Africa. These have a long history in trade of specialty woods and veneer logs, but otherwise they remain unexploited and lack access; or there is a lack of "know how" in utilisation of the intimate mixture of the many species which make up such forests.

The world's large, untapped softwood forests are in Canada and the USSR. Industry based on those in Canada will be of the greatest importance in determining trends and developments in international forest products trade for some time to come. Developments based on the Russian forests seem to be somewhat unpredictable. Transport problems are immense, and the financing and location of labour for the large-scale industries required seem to present difficulties not readily overcome. For all that, the resource is so large that in the long run it must provide great amounts of wood.

Until a few years ago, the heavily forested and relatively sparsely populated countries of Scandinavia and Canada were the biggest exporters of timber, pulp and paper. These countries dominated the world markets, but the former have almost reached the maximum in wood supply based on sustained yields and, although the Scandinavian countries still export large quantities of forest products, Europe as a whole has become a net importer.

Canada produces big amounts of timber, pulp and paper and in the past few years has become the world's only large net exporter. Her forest products industries are being expanded rapidly, and it has been calculated that her untapped forests could support a total production three to five times the present size. By far the largest market for Canada is the United States which, no doubt, will continue to take an increasing amount of Canada's expanding production. However, a recent national inventory has shown that the United States' wood resources are larger than had been realised. The United States is, therefore, expanding her own forest products industries considerably and will also look for some overseas markets. This in turn will force Canada to make additional efforts to find markets throughout the world for her increasing production; this is already happening and Canada is dominating more and more of the world trade.

New Zealand while building up a comparatively large afforestation programme (involving long-term use of land and finance) that looks forward to overseas markets for forest products has been very conscious of the need to consider prospects in relation to the world situation. Its experience so far has been that, provided exotic forests are grown in sufficiently large and concentrated areas in proximity to utilisation sites, the costs of wood delivered to these sites are comparable with the lowest in the world. This advantage should continue, although stumpages in New Zealand must rise to a level whereby growers are assured of profitable returns. By contrast, wood values in Europe are rising to high levels because of shortages, and in Canada the cost of transport is increasing as harvesting moves from the more accessible coastal forests to those of the interior.

On the manufacturing side, forest products industries are new to New Zealand. Although a great deal of progress has been made, there has been insufficient time to build up skills at all levels. Electricity and unit transport costs, both major items in pulp and paper manufacture, are also higher in New Zealand.

The value of the New Zealand forest products export trade will depend partly on the types of product marketed—the more advanced the processing is the greater the value will be as the following table shows.

FOREST PRODUCTS EXPORTS FOR YEAR ENDED
30 JUNE 1968

Commodity	Volume (million cu.ft)	Total Value f.o.b. ($000)	Return per cu.ft (cents)
Logs	43.2	11,750	27.2
Sawn Timber	13.0	4,747	36.5
Woodpulp	14.5	6,036	41.6
Newsprint (some other paper and board)	15.8	17,453	110.5
Sundry	0.2	1,235	—
	86.7	41,220	47.5

Since the highest values and the greatest hopes lie in pulp and paper, we must consider the world production of these commodities. Although our major market is likely to be Australia, that market must be affected by any change in the world situation.

World Trade in Pulp and Paper

World production and consumption figures of pulp and paper have been collated principally through the efforts of the Forest and Forest Products Division of the Food and Agricultural Organisation (FAO), United Nations. The division has also assessed consumption rates per head in all countries and has made predictions of consumption. Any country can therefore readily assess its position, present and predicted, from these world figures.

The most important feature of the whole industry is that it is expanding rapidly almost everywhere. Consumption of paper and paperboard has risen annually by 6 per cent over the past 10 years or more—from 42 million metric tons in 1949–51 to 75 million metric tons in 1959–61. Projected consumption for 1970 is 125 million metric tons, and for 1980, 203 million metric tons, or more than twice present consumption. Per capita consumption by main regions is as follows:

Per capita consumption of Paper and Paperboard

	1949–51 kg	1959–61 kg
North America	159.0	184.2
Latin America	8.5	12.9
Western Europe	31.2	60.5
Eastern Europe	13.3	23.2

USSR	—	—	—	—	—	8.0	15.2
Middle East		—	—	—	—	2.6	5.1
Near East	—	—	—	—	—	0.9	2.5
Africa	—	—	—	—	—	1.8	3.0
Asia (excluding mainland China)					—	1.9	6.4
China	—	—	—	—	—	0.5	4.0
Oceania	—	—	—	—	—	51.5	80.9

These figures reveal a remarkably wide variation in consumption per head between regions that are highly industrialised and those less so—over 700 per cent at the extremes. There are also differences between areas having extensive softwood forests and those with mixed hardwood forests, which are difficult to exploit for pulp production. As the effects of wider education are felt, and as industrialisation increases, the per capita consumption of paper in the less developed countries could rise faster than has been predicted. Demand is certainly always a great deal higher than supplies when these are limited by shortages of overseas exchange or shortages of raw material. Yet industrialisation often produces overseas earnings; and the range of raw material, particularly woods, being used for pulp and paper is continually expanding.

The supply and demand situation requires detailed study region by region. It is sufficient here to point out that the important figures to study are those for pulp production, because pulp is produced almost entirely by countries possessing the raw materials. There are two great centres of pulp production which have been traditional exporters of pulp and paper: North America and Scandinavia, which in 1966 produced almost 75 percent of the world's pulp. Although the total consumption in North America keeps increasing by huge amounts, raw material resources are sufficient to keep ahead of consumption and to provide for export for some time to come. New forest, particularly in Canada, is continually being opened up, and the growing stock in forest once cut over has been building up to amounts large enough for a second harvest.

All regions other than North America, Africa and USSR listed in the table are deficit areas in pulp production. Of special interest to New Zealand is that Oceania, which is mainly Australia and New Zealand, is a deficit area because of Australia's large deficit.

Regarding the world situation an FAO paper has recorded the following:

"... the world's requirements of pulp for paper and paperboard manufacture may rise from about 61 million tons in

1960 to 101.5 million tons in 1970 to 162 million tons in 1980. The prospective increase of over 100 million tons in two decades is impressive and gives reason for pre-occupation about the distribution and allocation of fibre resources, since it is unlikely, indeed, that the traditionally net exporting regions of North America and Western Europe could be called upon to satisfy more than a small share, if any, of the growing needs of the deficit areas. Nor is it likely, for that matter, that the net importing regions could afford to substantially raise their imports from the industrialised regions."
This quotation refers, no doubt, to the later part of the period with which we have dealt.

It appears that before long, North America must become the only large pulp-exporting region, and for this reason a good deal of attention is being focussed in the deficit areas on the breeding or discovering of fast-growing trees for the production of pulpwood. In many parts of Europe poplars are being grown; but the greatest prospects of producing large supplies lie in fast-growing exotic conifers in the warm and temperate regions of the more recently settled countries where adequate planting land is available. Africa, for example, has a combination of suitable land and introduced conifers, as well as other trees. She could grow raw material for her own pulp requirements and also look to the possibilities later of exporting to the deficit countries of Europe. Indeed, the development of exotic afforestation on a large scale offers some African countries encouraging possibilities of raising standards above a subsistance agriculture and of earning overseas exchange. The same sort of potential exists in many parts of South America. A recent FAO statement is introduced with the following: "A realistic program of national forestry development, under which Chile might in comparatively few years become the Sweden or Denmark of Latin America, was formally proposed yesterday to the Government of Chile by the Food and Agriculture Organisation."

New Zealand is particularly favoured in forest potential, and is ahead of most countries in the creation and management of fast-growing exotic forests and in the use of these for pulp and paper for internal and export markets. The stage is set for capitilising on this position and taking advantage of growing world shortages to expand exports of pulp and paper. Forestry in New Zealand has the same main advantage of its agriculture, a very favourable growing climate; but whereas industries based on agricultural produce are not highly developed, some of those based on wood are. To make the most of this situation New Zealand must pay special attention to research and to training at **all levels.**

8

TRAINING AND RESEARCH

Forests in the proximity of ancient civilisations were cut for use without thought of replacement. In some regions this destruction in turn undoubtedly had adverse effects on the civilisations concerned because it deprived succeeding generations of wood supplies to which the community as a whole had become accustomed.

Management practices aimed at perpetuating wood supplies developed very slowly in marked contrast to the progress of agriculture. By Greek and Roman times there was some knowledge of coppicing—the regrowth of a tree or shrub from the stump—and such a system was used for obtaining, particularly, perpetual supplies of firewood but not of timber. There was also some knowledge of raising tree seedlings from seed and planting the seedlings out. This and whatever other forestry knowledge was built up found its way to the Roman-occupied countries. In the ancient German countries such knowledge took strong root and continued to develop until Germany was recognised as having the best forest practices in the world and some of the best-managed forests. These developments, which took several centuries, depended mainly on an understanding of the manner in which forests grow; but important complementary developments were the evolution of satisfactory land tenures, the passing of laws regulating the felling in and management of State and private forests, and so forth.

Modern development of forest practice did not occur until it was possible to measure accurately areas of land and volumes and increment growth of trees and whole stands of trees. Until these things had been done, yields could not be regulated accurately and the effect of thinning and other practices on stands could not be gauged in terms of growth of wood.

By the eighteenth century forest services were fairly well organised and some of the master foresters, the foremost thinkers of the day in forestry matters, organised private schools. A few of these later developed into State schools and later still into schools for the teaching of forestry at university level. About the same time forest experimental stations were established in Germany. These two developments, the experimental approach to forestry and the teaching of it at universities, laid the true foundation of modern forest practice. German teaching soon spread in Europe and to many countries round the world.

A renowned German forester and teacher, William Schlich, who had served in the British Colonial Forest Service and was Inspector General of Forests for India, was appointed in 1885 to the first forestry school in the British Empire—Coopers Hill. This was attached to the Coopers Hill School of Engineering of the University of Oxford. Foresters trained there went on to develop the most successful British Colonial Forest Service, the best-known branch of which was the Indian Forest Service. It will be remembered that Captain Inches Campbell Walker of that service was invited to New Zealand by the Prime Minister, Julius Vogel, to advise him on forestry matters and to help draft the first Forests Act of 1874.

Modern forestry, therefore, is a comparatively recent development which was scarcely under way when settlement of New Zealand by Europeans was beginning. Early immigrants could have had neither general forestry knowledge nor ideas of how to apply it to the management of New Zealand's complex forests, even had they thought this necessary in the face of huge surpluses of standing timber. This makes all the more remarkable, therefore, the passing of the first Forests Act early in 1874 and the provision in the second (1885) Act for setting up of a school of forestry at a time when German university schools were only just becoming well established. True, the forestry school did not function because the general climate of opinion was not ready for it, but from that time the idea of it was never wholly dropped.

The actual setting up of a forestry school seems, however, not to have been mooted officially again until the first Director of Forestry proposed it in his report on forest policy to the House of Representatives in 1920. By this time two strong influences that had developed as New Zealand forestry progressed made themselves felt. One arose as a result of the earlier importance of the kauri forests. This importance had become ingrained in thinking of that time, although the forests had been virtually cut out and their value greatly diminished. The other influence came from the originally treeless Canterbury plains, on which early plantings of shelter belts and woodlots had represented the beginnings of exotic forestry.

These two strong influences found voice through the parochial attitudes of the Auckland and Canterbury Colleges of the University of New Zealand. Both laid claim so strongly to the proposed forestry school that the Government finally compromised and schools were opened at both colleges—Canterbury in 1925 and Auckland in 1926. Both were too small and poorly staffed and equipped for really efficient teaching of forestry subjects, although basic science subjects were covered adequately in other faculties.

The undesirability of having two schools was realised at the beginning and, after a report by two eminent Empire foresters who had attended an Empire Forestry Conference held in New Zealand in 1928, the Auckland school was closed in 1930. There followed years of severe recession, which led to setbacks in all Government activities. During this period it was proposed that the newly created State Forest Service should be reunited with the Department of Lands and Survey: but fortunately for the progress of forestry its importance was not written down in this way. Regrettably, though, the retrenched Forest Service could not offer positions to graduates from the schools of forestry, and in 1934 the Canterbury school also was closed, partly because of lack of employment opportunities and partly because of cuts in Government expenditure. Scarcely 20 students graduated from the two short-lived schools.

The only forestry practised in this country for a considerable time was exotic forestry developed from the enthusiasm for plant introduction and the success of it. The Afforestation Branch of the Lands and Survey Department was staffed by skilled nurserymen and men knowledgeable about trees but not about forests. A few were given the opportunity of inspecting forests and forest practice in Europe, and these men laid a very sound foundation for the raising and planting out of introduced trees. Among local bodies and private owners in this development, the foremost were the Canterbury counties, the forerunners of the Selwyn Plantation Board, who sent their chief officer abroad for experience.

Not until the State Forest Service was formed were professional men recruited to guide forestry conceived on a wider basis than just afforestation with exotic trees. Nevertheless, only a handful of men provided this guidance for the first 20 years after the formation of the State Forest Service and only one experienced professional forester was employed by private enterprise throughout the 1925–35 planting boom.

In 1939 a newly appointed Director of Forestry, A. R. Entrican, visualised the forestry potential of this country, and saw that the correct way to develop this was to train personnel at all levels. Within a few years, therefore, the Forest Service began training schemes for woodsmen (who were to be the skilled forest workers and forest tradesmen of the service), for forest rangers (who were the chief field officers), and for professional foresters. It was tacitly accepted by the Government that some of these trained men would be enticed away from the service by private enterprise. In fact, more of these officers, especially forest rangers, have found their way into private enterprise than have remained with the Forest Service.

Because there was now no forestry school in New Zealand, men selected for professional training obtained their basic science degree in this country and then went to overseas forestry schools to train in forestry. Most have gone to the Australian Forestry School, Canberra, but a proportion has also gone to British forestry schools at Oxford, Edinburgh, Aberdeen, and Bangor (Wales). A few officers, usually for training in specialised fields such as logging, have been sent to North American schools, and two, who have mastered the necessary languages, to European schools. Altogether, professional forestry officers in New Zealand have been trained at nearly 20 forestry schools in various parts of the world.

This system of training has produced a cadre of officers within the service who have a broad approach to the practice of forestry. Nevertheless, the wish to have a school of forestry in New Zealand has persisted in some quarters and as a result of strong pressures recently the Government agreed to the re-establishment of a school on the Ilam campus of the University of Canterbury.

At Ilam the new school will be associated with a Forest Service research establishment for the South Island, which the university authorities have agreed will be located there. It will also be close to a school of engineering, to Lincoln Agricultural College, and to a number of other university units which should stimulate forestry teaching. Few centres anywhere in the world have the same advantages.

Thus the legislation enacted over 80 years ago will at last result in a permanent forestry school. The stimulus to New Zealand forestry should be considerable, but it is desirable that a proportion of post-graduate training should still be taken overseas.

A sawmill cutting native timber. The shanty building, very poor surroundings, and poor equipment and accommodation are typical of a transient industry based on short-term cutting rights. Mills are abandoned as timber is cut out, and usually agriculture takes over the land that has been cut over. This limited-life industry produced New Zealand's timber requirement for about a hundred years.

(Page 76)

Japanese ships loading radiata pine logs at Mount Maunganui, Bay of Plenty. This port, originally constructed almost solely for the export of pulp, paper, and timber, has been developed rapidly to cope with expanding log exports. Total forest

Forestry and Forest Products Research

In its first annual report (1965) the National Research Advisory Council (established in 1963 by Act of Parliament to advise a Minister of Science on scientific research) has this to say about forestry research:

"Foresty has a unique position in any country's economy because wood is one of the major natural resources that is renewable. This is especially significant in New Zealand, where the growth rate of pines is extraordinarily high and where, after only 40 years of serious planting, our manmade forests have a monetary value greater than that of our usable native forests. New Zealand has already some £370 ($740) million invested in standing timber. The growth of the forest-based industries is illustrated by the rise in value of output from £76.6 ($153.2) million in 1959–60 to £93.2 ($186.4) million in 1962–63."

Concerning forest products research the report says:

"Research to improve timber quality should have a high priority in view of the increased need to use exotic timbers for finishing grades in face of the diminishing supplies of native timbers. In the field of pulp and paper production the research effort to date has been restricted primarily to that required by industry to establish commercial operations. Little effort has been made to study the details of the processes or even to examine the suitability of raw materials other than those represented by the main crop of the exotic forests. There has also been little research by Government agencies in the field of pulp and paper."

The council recommended an immediate increase in forestry and forest products research. This recommendation almost coincided with the calling of tenders for the first stage of new buildings for the Forest Research Institute, Rotorua, the first permanent ones for the Institute.

As long as ample indigenous timber for building and construction was available from forests that were being cleared for settlement, there was little justification for expenditure on research. As far as knowledge of timbers was concerned, settlers began at the stage which the Maoris had reached and they continued to develop that knowledge from practical experience. No attempt was made to introduce forestry practice into the indigenous forests with the aim of perpetuating supplies. Botanists did set to work to identify plants, including forest trees, but this was scarcely research in the forestry sense. Nevertheless botanists were among the first technical people to be closely associated with forestry questions. Thomas Kirk, a noted botanist, reported on forests throughout the country and also wrote a forest *Flora* in which are described many native trees and their timbers. The timber descriptions were, however, very general and were mainly compendiums of uses to which timbers had already been put.

The first botanist, this time an ecologist, to describe forests scientifically was Leonard Cockayne, who for a short time was secretary to the 1913 Royal Commission on Forestry in New Zealand. He was the first person to classify forests and to produce hypotheses of their origins and life histories. Although not a taxonomist, he discovered the amount of natural hybridisation that takes place in native plants, including forest trees—hybridisation that can be detected in the wood as well as in the form and structure of trees. A prodigious worker, Cockayne produced an extensive volume of ecological work of world class. As one of the principal scientists to foster the study of modern plant ecology he greatly influenced the growth of forest ecology in New Zealand and this development has been continuously refined and changed as new knowledge has come to light. Such ecological work is of the greatest importance as the basis to most native forest management, whether management of kauri, beech, or rimu forests for production or the management of protection forests to safeguard them from the depredations of animals. New Zealand foresters owe a great deal to Cockayne.

With the new Forests Act and new department in 1919-21 came the new policy, which included recommendations for the establishment of "A Forest Products Laboratory and Bureau of Forest Research, survey and inventory of the forests, forest resources, and soils of the Dominion". These recommendations were acted on to a limited extent only, a forest products research section being set up at the head office of the Forest Service in the Dominion Farmers' Institute Building, Wellington. Some forest products research and forestry research was done at the laboratories of the University of New Zealand's engineering and forestry schools in Auckland and Christchurch, at the Cawthron Institute, Nelson, and at the Department of Scientific and Industrial Research laboratories, Palmerston North. Most emphasis was on forest products research, and the empirical work of the past was replaced by scientific experimentation. Investigations were made into timber properties, seasoning and preservation; and pulp and paper trials on some native woods (including rimu) and on radiata pine were carried out in overseas laboratories. These trials, although on a very modest scale, paved the way for the pulp and paper industry many years later. Investigations into the properties of exotic timbers were increased as the time approached to saw and market them. This research was an essential stage in developing the utilisation of such timbers.

Forest products research by the State—principally by the Forest Service—on the basic properties of wood, on seasoning, and on preservation has a good record, but very little or nothing has been done on the utilisation of veneers and plywoods, chipboards, finger jointing, wood-based boards, and other composite products, and little in pulp and paper products. These are the commodities that industry has developed or is moving into and research, in the absence of effort by the State, has developed on a company basis. Research has sometimes preceded the development of a commodity, especially when it is to be produced by a large firm, but frequently it has followed production and only when difficulties have begun to appear. Because the State has a great deal at stake in the efficient use of wood sold to produce some of these commodities, there is every reason for it to extend forest products research into this field. The first wing of the new Forest Research Institute is for the forest products research section and it will provide enough laboratory space to permit the needed new developments.

The close association of forestry and forest products research —most State research is carried out in the one research establishment—enables scientists working on products to examine with foresters the natural variation of wood found in individual trees, in the trees of a single stand, and in stands throughout the country. They can study the effects on wood of different thinning intensities, of pruning and of other treatments. It is equally important for the forester to understand what effects his forest treatments are having on wood qualities. The selection of improved trees—as far as these improvements affect wood properties—can be examined and guided by forest products research workers. For these and other reasons it is important that this close association, already developed in New Zealand, be maintained.

The forestry research that had been begun in the two university schools and elsewhere was finally gathered together and with forest products research was housed in an experimental station at Whakarewarewa, Rotorua, in 1947. A small group of scientists moved into old buildings which had been used as stables and offices during the operation of the Whakarewarewa forest nursery and were later used for the State Forest Service Rotorua Conservancy.

A much more active phase of forestry research now began. Projects started in the schools and various places over 20 years previously had largely been dropped. Sample-plot and other mensuration work in the growing of exotic forests had been carried out in a desultory manner and some pathological work had been undertaken, mainly in connection with diseases appearing from time to time in the exotic forests. The new vigour came partly from the beginning in 1946 of a National Forest Survey of native forests; it was completed in 1950.

After this survey, two others were begun: one was a survey of protection forests and other protective vegetation interspersed with forests on the mountains; the other, an exotic forest survey. The protection forest survey was based on Rangiora in the South Island and, although it was primarily a survey, staff commenced to undertake research to solve some of the many problems which survey work soon revealed. The South Island unit—known first as the Forest and Range Experiment Station and now as the Protection Forestry Division of the Forest Research Institute—began when the Forest Service took over noxious animal control and it has been concerned with the important work of assessing the condition of protective vegetation and understanding the changes going on in it.

The remarkable activity near the end of last century and the beginning of this century in introducing and acclimatising animals and the expenditure incurred were not matched at all by research. It is difficult to understand why, because while rabbit populations were at their peak, their depredations cost agricultural producers hundreds of millions of dollars and hindered much technical progress. Rabbits and larger animals seriously depleted millions of acres of protective vegetation. Studies on the larger animals were started about 1950 and are being increased in the Protection Forestry Division and in the Department of Scientific and Industrial Research. Scientists must be young and exceptionally fit to undertake many of the studies in the difficult terrain and rigorous climate.

Tree introduction, which began over 100 years ago, and the results of which are to be seen in the present day exotic forests, was undertaken on an increased scale and on a scientific basis by the Forest Research Institute. Many new species and variations of those already grown here have been introduced. For example, seed from the natural radiata pine stands in California has been collected and seed from stands of what are considered to be radiata stands on islands off the coasts of California and Mexico.

The large-scale use of exotic wood has produced many silvicultural and other forest-management problems, and some of these can be solved only with the help of research. Moreover, within a single decade, the position has changed from one in which industry had on call large surpluses of exotic wood to one in which there are already shortages in some districts. Where there are still surpluses, industry has plans for their utilisation. As the accurate measurement of volumes of standing timber and increments of stands is becoming increasingly important research in these fields is being expanded.

Forestry is usually expected to use land that is difficult to develop, often carrying luxuriant and useless second-growth of native species or infested with gorse or other weeds. The preparation of such ground for planting presents many difficulties because normal cultivating implements cannot be used. Desiccants sprayed from the air followed by burning are about the only "tools" at the forester's command. The competent use of both these, on the extensive scale required, involves research.

The working of exotic forests has brought foresters face to face with hitherto unsuspected silvicultural and management problems. Intensities of thinning affect the rates of growth of individual trees and of whole stands, log sizes and quality of wood. These in turn all affect the economics of wood production. Heavy thinnings produce logs more economically than do light thinnings, and the trees left behind grow more quickly because they have more space, but wood quality might not be so good because it will be coarser. A host of individual and interrelated problems arise which will depend for a solution on adequate research.

In native forests coming under management for sustained yields similar problems require attention, but are more difficult to solve than are those in exotic forests because the forests are more complex and growth is much slower. Results of experiments, therefore, take longer to appear. In these circumstances the continuity of measurements and observations provided for in a research institute becomes of great importance. In fact this need for continuity, often throughout long periods, is the most difficult aspect of forestry research. Often two or more observers must carry through the one experiment, and techniques almost always change during a long experiment.

Although forestry is usually relegated to land unsuited for agriculture, we have reached a stage in the development of both when their economics must be closely examined in relation to land use. Land that can be developed easily for agriculture is becoming difficult to acquire, and forestry is making strong demands for some that remains because of the raw material needed for expanding forest products industries. Very intricate economic research into land use therefore becomes desirable, and such investigations have been started both by agriculture and by forestry.

Like all crops, forests are subject to diseases. Usually the more intensive the cultivation is the greater the risk becomes. In unmanaged native forests attacks on trees usually pass unnoticed, unless particularly severe, because they are economically unimportant; in a managed or utilised forest, however, they are soon noticed because they involve loss of money. Trees in native forests are seldom subject to epidemic diseases, though from time to time beech forests are extensively attacked by a beetle (*Nascioides*) which kills scattered trees over wide areas. This beetle has damaged small areas of managed beech stands. In the last few years, a disease, as yet unidentified, has been killing totara in forest on the edge of the Urewera.

From time to time diseases strike exotic forests. These attacks could be much more widespread and frequent if there was no quarantine service at ports and aerodromes, but some diseases or pests inevitably escape and others are borne by wind from overseas.

Introduced Australian insects have damaged extensively many *Eucalyptus* species, and a European wood wasp (*Sirex*), has severely attacked radiata pine, fortunately at a time when considerably more wood was available than industry was using. *Dothistroma pini*, a needlecast fungous disease of pines, is the most threatening disease to have appeared so far. An indigenous fungus of North America it has spread to Europe, evidently without causing damage, and to the highlands of East Africa, where it has infected radiata pine severely. During the past few years in New Zealand it has spread rapidly through young radiata pine stands (it does not seem to attack older stands of this species) in the central North Island, and has appeared in nurseries and stands of this species elsewhere. It has also attacked stands of pondersosa and Corsican pines of all ages. Copper sprays control it for a time and the cost of them can be borne by the fast-growing, short-rotation radiata pine.

It is too early to assess the ultimate severity of this disease or the effect it will have on management and silviculture. Research is unravelling the life history and is throwing light on many other facets of the disease and its control.

A highly skilled and experienced team of research workers is required—and available—for work on problems such as these.

APPENDIX I

PRINCIPAL NEW ZEALAND ACTS AFFECTING FORESTRY

Forest Trees Planting Encouragement Act 1871
Forest Trees Planting Encouragement Act Amendment Act 1872
New Zealand Forests Act 1874
Land Act 1877
Forest Trees Planting Encouragement Act Amendment Act 1879
Land Act 1877 Amendment Act 1882
New Zealand State Forests Act 1885
Land Act 1885
New Zealand State Forests Amendment Act 1888
Land Act 1892
State Forests Act 1908
Land Act 1908
Forests Act 1921–22
Land Act 1924
Forests Amendment Act 1925
Forests Amendment Act 1926
Statutes Amendment Act 1939
Statutes Amendment Act 1941
Forest and Rural Fires Act 1947
Forests Act 1949
Forest and Rural Fires Act 1955
Noxious Animals Act 1956
Forestry Encouragement Act 1962

APPENDIX II

TIMBER TREES NAMED IN THE TEXT

Native Softwoods

Kahikatea	*Podocarpus dacrydioides*
Pahautea	*Libocedrus bidwillii*
Kauri	*Agathis australis*
Matai	*Podocarpus spicatus*
Rimu	*Dacrydium cupressinum*
Silver pine	*Dacrydium colensoi*
Tanekaha	*Phyllocladus alpinus*
Totara	*Podocarpus totara*

Introduced (exotic) Softwoods

Douglas fir	*Pseudotsuga menziesii*
Japanese cedar	*Cryptomeria japonica*
Larch—European	*Larix decidua*
Larch—Japanese	*Larix leptolepis*
Lawson cypress	*Chamaecyparis lawsoniana*
Macrocarpa	*Cupressus macrocarpa*
Mexican cypress	*Cupressus lusitanica*
Pine—Austrian	*Pinus nigra* (austriaca)
Pine—Corsican	*Pinus nigra* (laricio)
Pine—loblolly	*Pinus taeda*
Pine—lodgepole	*Pinus contorta*
Pine—maritime	*Pinus pinaster*
Pine—patula	*Pinus patula*
Pine—ponderosa	*Pinus pondersosa*
Pine—radiata	*Pinus radiata*
Pine—Scots	*Pinus sylvestris*
Pine—slash	*Pinus elliottii*

Pine—strobus	*Pinus strobus*
Redwood	*Sequoia sempervirens*
Western hemlock	*Tsuga heterophylla*
Western red cedar	*Thuja plicata*

Native Hardwoods

Beech—black	*Nothofagus solandri* var. *solandri*
Beech—hard	*Nothofagus truncata*
Beech—mountain	*Nothofagus solandri* var. *cliffortioides*
Beech—red	*Nothofagus fusca*
Beech—silver	*Nothofagus menziesii*
Rata	*Metrosideros robusta*
Rewarewa	*Knightia excelsa*
Tawa	*Beilschmiedia tawa*

Introduced (exotic) Hardwoods

Blue gum	*Eucalyptus globulus*
Callitris	*Callitris* spp.
Catalpa	*Catalpa speciosa*
Crack willow	*Salix fragilis*
European ash	*Fraxinus excelsior*
Oak	*Quercus robur*
Poplar	*Populus* spp.
Turpentine	*Syncarpia laurifolia*
Wattle—silver	*Acacia dealbata*
Wattle—black	*Acacia decurrens*